The Development of the B-52 and Jet Propulsion

A Case Study in Organizational Innovation

DR. MARK D. MANDELES

Air University Press
Maxwell Air Force Base, Alabama

March 1998

Library of Congress Cataloging-in-Publication Data

Mandeles, Mark David, 1950-
The development of the B-52 and jet propulsion : a case study in organizational innovation /
Mark D. Mandeles.
p. cm.
Includes bibliographical references and index.
  1. Aeronautics, Military-Research-United States-History. 2. B-52 bomber- Research- History.
3. Jet propulsion-History.
I. Title.

UG643.M35 1998   98-14703
358.4070973-dc21   CIP

Printed March 1998

Digitize Copy from March 1998 Printing November 2002 NOTE: Pagination changed

DISCLAIMER

This publication was produced in the Department of Defense school environment in the interest of academic freedom and the advancement of national defense-related concepts. The views expressed in this publication are those of the authors and do not reflect the official policy or position of the Department of Defense or the United States government.

This publication has been reviewed by security and policy review authorities and is cleared for public release.

From Model 462 . . .

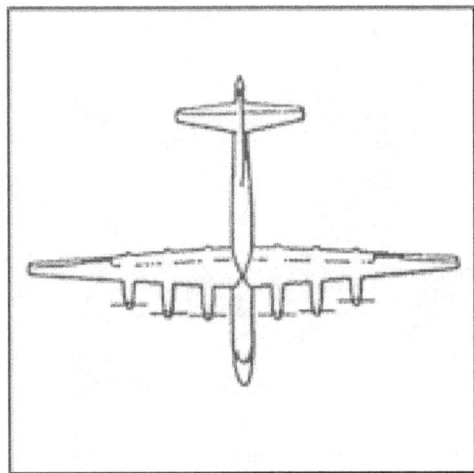

to Model 464-49 . . .

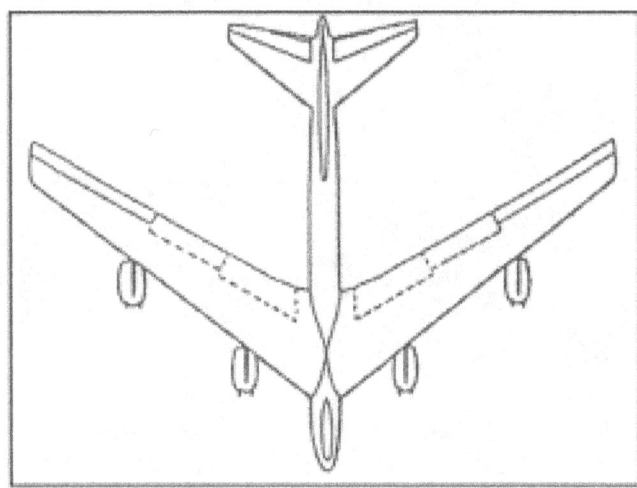

to the B-52.

Contents

Foreword

The B-52 and Jet Propulsion: A Case Study in Organizational Innovation is a coherent and nonpolemical discussion of the revolution in military affairs, a hot topic in the national security arena. Mark Mandeles examines an interesting topic, how can the military better understand, manage, and evaluate technological development programs. We see Murphy's Law (anything that can go wrong, will go wrong) in operation. No matter how carefully the military designs, plans, and programs the process of technological development, inevitably, equipment, organizations, and people will challenge the desired expectations. Mandeles argues convincingly that recognizing the inevitability of error may be the single most important factor in the design of effective organizations and procedures to foster and enhance innovative technology and concepts.

The book focuses on the introduction of jet propulsion into the B-52. This case study illustrates the reality that surprises and failures are endemic to development programs where information and knowledge are indeterminate, ambiguous, and imperfect. Mandeles' choice of the B-52 to illustrate this process is both intriguing and apt. The military had no coherent search process inevitably leading to the choice of a particular technology; nor was decision making concerning the B-52 development program coherent or orderly. Different mixtures of participants, problems, and solutions came together at various times to make decisions about funding or to review the status of performance projections and requirements.

Three aspects of the B-52's history are striking because they challenge conventional wisdom about rationally managed innovation. First, Air Force personnel working on the B-52 program did not obtain the aircraft they assumed they would get when the program began. Second, the development process did not conform to idealized features of a rational program. While a rationally organized program has clear goals, adequate information, and well-organized and attentive leadership, the B-52 development process exhibited substantial disagreement over, and revision of, requirements or goals, and ambiguous, imperfect, and changing information. Third, the "messy" development process, as described in the book, forestalled premature closure on a particular design and spurred learning and the continuous introduction of new knowledge into the design as the process went along.

Military innovations involve questions about politics, cooperation and coordination, and social benefits, and like other development efforts, there appears to be no error-free method to predict at the outset the end results of any given program. This study offers a major lesson to today's planners: improving the capacity of a number of organizations with overlapping jurisdictions to interact enhances prospects to innovate new weapons and operational concepts. We can mitigate bureaucratic pathologies by fostering interaction among government and private organizations.

The B-52 and Jet Propulsion integrates a detailed historical case study with a fine understanding of the literature on organization and innovation. It is a story of decision making under conditions of uncertainty, ambiguity, and disagreement. I have seen such stories unfold many times in my work on technological development projects. In the pages that follow those who plan, manage, and criticize technological development programs will find new insights about the process of learning how to make new things.

DOV S. ZAKHEIM
CEO, SPC International Corporation

About the Author

Mark D. Mandeles won research and university fellowships from the University of California, Davis, and Indiana University. In 1982 he won the USAF Dissertation Fellowship in Military Aerospace History, and subsequently received a doctor of philosophy degree in political science from Indiana University. He also earned a master of arts degree in political science from the University of California, Davis, and a bachelor of arts degree in political science (with honors) from the University of California, Berkeley.

For the last 15 years, he has been an analyst or consultant for government and private analytical firms, including the Director of Net Assessment in the Office of the Secretary of Defense, the Secretary of the Air Force's Gulf War Air Power Survey, the Air Staff (Office of Long-Range Planning), Institute for Defense Analyses, Center for Naval Analyses, US General Accounting Office, Office of Chief of Engineers (US Army Corps of Engineers), Center for Air Force History, Science Applications International Corporation, OC Inc., ANSER, and Delex Systems. He founded the J. de Bloch Group, a research and analysis firm, in 1993. Currently he is analyzing structures of future military organizations for the Office of Net Assessment.

His published research and reviews (in Security Studies, Naval War College Review, American Political Science Review, National Defense, Military Affairs, Journal of America's Military Past, and Middle East Insight) include essays on the revolution in military affairs, weapons acquisition, innovation in naval aviation, and ballistic missile and nuclear weapons proliferation. He coauthored Managing "Command and Control" in the Persian Gulf War (Praeger, 1996), and The Introduction of Carrier Aviation into the U.S. Navy and Royal Navy (Naval Institute Press, forthcoming).

Dr. Mandeles lives in Alexandria, Virginia, with his wife, Laura, and children, Samantha and Harry.

Acknowledgments

I have had outstanding help from knowledgeable, kind, and gracious colleagues in preparing this manuscript, including Thomas C. Hone, Laura L. Mandeles, Andrew W. Marshall, Jacob Neufeld, and Jan M. van Tol. Of course, I alone am responsible for any errors they were unsuccessful in persuading me to remove. I also wish to thank the wonderful staff of the John Marshall branch of the Fairfax County (Virginia) Public Library for securing many books through interlibrary loan and Col Allan W. Howey, director of Air University Press, and the AU Press editorial team, Emily Adams and Lula Barnes.

To
Martin Landau
and
Andrew W. Marshell

Chapter 1
Introduction

When asked why he had not submitted more work for publication, Henry Roth, the author of the critically acclaimed 1934 novel, Call It Sleep, responded, "If I had been in a more stable society, one that hadn't changed so abruptly (due to upheavals of the Great Depression and World War II), I could have gotten-oh, like Dickens. He could count on his society, and his attitude toward that society, being the same from his first novel to the last. I didn't feel I could do that." [1]

Like Roth, national security decision makers face an uncertain world where the accelerated growth of knowledge has changed the character of technological advance and destabilized long-standing relations within and among the military services. But unlike Roth, national security decision makers do not have the luxury of withdrawal; they must try to make reasonable judgments about acquisition of military technology, associated concepts of operations, and the organization of combat forces.

State of the art technology offers military planners a dazzling array of advanced weapons systems, communications equipment, computer tools, and more. Indeed, in the aftermath of the Persian Gulf War, some military analysts have even hailed the possible emergence of a "military" revolution-an order of magnitude increase in combat capability. A military revolution occurs when a set of technologies and associated operational concepts transform the nature and character of warfare, and military organizations and their personnel are able to deploy and exploit the set of technologies. These observers have noted the Gulf War performance of stealth aircraft, precise long-range conventional munitions, and advanced sensor, targeting, and information processing technology and have suggested that major improvements in combat effectiveness are impending as these technologies are integrated into military forces. [2]

Yet, with great opportunities come great uncertainties. Little consensus exists about the comparative payoffs-the combat effectiveness-of each service's weapons acquisition programs. Rapid development of military technology presents difficult and complex problems of choice and of how to organize to make those choices effectively. What factors should be considered and how should choices be made about new programs or continued system development? How should new equipment be integrated into existing combat organizations and concepts of operations? What criteria indicate that existing combat organization, operational concepts, and weapons should be abandoned, separately or altogether? What bureaucratic and organizational processes obstruct thorough peacetime consideration of the effects and opportunities of advancing military technology?

Existing decision processes and organizational structures for military acquisition too often have led to cost overruns, schedule delays, and performance shortfalls. The task of supporting useful development and discouraging failure challenges policy makers to design and implement new organizational forms appropriate to fostering innovation and emerging mixtures of weapons, operational concepts, and skills. Yet, we have little "hard" knowledge about how to organize to advance innovation or to adapt to rapidly changing circumstances. Military decision makers face a situation neatly captured years ago by reform journalist and humorist Finley Peter Dunne's character, Mr. Dooley: "It ain't what we don't know that bothers me so much; it's all the things we do know that ain't so." Policy analysts assume knowledge about how to encourage innovation where there is none and tend to assume casual relationships and the importance of particular variables on the basis of very little evidence. Systematic comparisons and analyses have not been conducted of the critical relationship between organizational structures (e.g. hierarchy and types of administrative redundancy and overlap and organizational outcomes. Consequently, what frequently passes for "principles" of organization or design are really "proverbs." [4]

The use of principles to guide decision-making permits more self-conscious evaluation of causes and effects. This monograph seeks to separate the principles from the proverbs through a case study of decision-making in the early post-World War II period. The case study examines the impact of organization on the invention and development of jet propulsion-in the form of the B-52.

The effort to develop this new weapon system began in a peacetime period of great uncertainty as senior military leaders struggled to anticipate the nature of future wars and to understand the implications of new technologies for existing roles and missions. Many strands of activity converged in the development of a long-range bombardment aircraft in the late 1940s and early 1950s: (1) the potential military threat posed by the Soviet Union, (2) bureaucratic battles over the budget, (3) armed forces unification, (4) disagreement over missions and roles, (5) the evolution of Air Force and Navy strategic doctrine, and (6) the employment of atomic weapons. How civilian and military leaders went about the task of conceiving and considering options provides interesting history. But even more important it provides the opportunity to examine the organizational structures in which these systems were developed-structures, which were among the most conducive to innovation in the history of the American military. [5]

The case study illustrates both the organizational conditions conducive to developing new operational concepts and the organizational innovations necessary to implement new technology. The study examines how the Air Force organized to learn and acquire new technology, how the Air Force conceived or identified problems, and how it organized to ensure management would respond to program failure or errors. In particular, attention is devoted to the origins of the weapon system operational requirement, the initial concept of operation, and the evolution of technology, organizational structure, and implementation.

In military innovation especially, the effort, time, attention, resources, expense, and good luck required to introduce innovations routinely into organizations involve a complex set of interactions creating opportunities for error, delaying implementation, or even dooming the innovation. Military innovations involve questions about policies, cooperation and coordination, and such social benefits as who derives status and authority from new operational concepts of weapons, what are the costs of change and who pays those costs, what opportunities for administration, management, and promotion are afforded to those dedicated to new programs, and how trust develops among people dealing with the innovation. [6] Hence, it is not surprising that military innovations take years to become routine. In Stephen P. Rosen's survey of important twentieth century military innovations, roughly a generation passed before they became an operating, routine, and integral part of military organization. [7]

Military innovation usually is examined via case studies. To be useful, the case study method requires a context or theoretical understanding that links the particular cases. Chapters 2 and 3 of this book provide the reader with the context. Chapter 2 identifies some general principles to guide policy makers as they try to anticipate a military revolution. Chapter 3 shows how understanding multiple levels (or units) of analysis is critical to an understanding of military innovation. Chapters 4 and 5 detail the introduction of jet propulsion into the B-52.

Notes

1. Quoted in David Stratified, "Bookends of a Life." The Washington Post, 15 February 1994, E6.

2. See, for example, John W. Bodnar, "The Military Technical Revolution: From Hardware to Information," Naval War College Review 46, no. 3 (Summer 1993): 7-21: Ashton B. Carter, William J. Perry, and John D. Steinbruner, A New Concept of Cooperative Security, Brookings Occasional Papers (Washington, D.C.: The Brookings Institution, 1992), 3, 29-30: James R. FitzSimonds and Jan M. van Tol. "Revolutions in Military Affairs," Joint Force Quarterly, Spring 1994, 24-31; Dan Gouré. "Is there a Military-Technical Revolution in America's Future?" The Washington Quarterly 16, no. 4 (Autumn 1993): 179: Richard P. Hallion, Storm Over Iraq: Air Power and the Gulf War (Washington, D.C.: Smithsonian Institution Press, 1992); and William J. Perry, "Desert Storm and Deterrence," Foreign Affairs 70 (Fall 1991): 66.

3. Richard R. Nelson, "A Retrospective," in Nelson, ed., National Innovation Systems: A Comparative Analysis (Oxford: Oxford University Press, 1993), 505. See also Charles E. Lindblom and David K. Cohen, Useable Knowledge: Social Science and Social Problem Solving (New Haven: Yale University Press, 1979).

4. This situation has changed little since Herbert A. Simon's essay, The Proverbs of Administration," Public Administration Review 6, no. 1 (Winter 1946): 53-67.

5. Harvey M. Sapolsky argues that the organizational structures most conducive to military innovation existed during the 1950s. He omits explicit reference or comparison to the interwar US Navy's organizational and institutional structures that related to aviation. 1n the absence of an explicit comparison, it is premature to argue that the organizations and institutions of the 1950s were superior. See Sapolsky. "Notes on Military Innovation: The Importance of Organizational Structure," 24 February 1994, unpublished paper. 5.

6. Arthur L. Stinchcombe, Information and Organizations (Berkeley: University of California Press, 1990), 155-76.

7. Stephen P. Rosen, Winning the Next War: Innovation and the Modern Military (Ithaca: Cornell University Press. 1991), 259.

Chapter 2
Innovation and Military Revolutions

Where observation is concerned, chance favors only the prepared mind.

—Louis Pasteur

A dictionary definition of innovation is "effecting a change in the established order; introduction of something new, [and] the change made by innovating; any custom, manner-newly introduced."[1] The word entered the English language in the sixteenth century from French and Latin roots on the eve of far-reaching social and political changes spurred by the Industrial Revolution.[2]

Innovation has always been a part of human existence; it is the result of curiosity, the incremental elimination of defects, and the effort to find a better way.[3] Ideas for technological innovation often emerge from the users of a technology.[4] However, not every change qualifies as an innovation. Innovation, in other words, refers to new things and new ways of carrying out tasks.[5]

Accelerated Growth of Knowledge

Before the twentieth century, the rate of military and civil invention was slower and the time elapsed during diffusion and deployment was greater than today. There were relatively fewer fundamental changes in the character of warfare. Few innovations occurred, for example, between the battle of Breitenfeld in 1631 and Waterloo almost two centuries later.[6] A soldier could learn the tools and tactics of a military career and not expect any drastic technology-driven changes over a lifetime.

The current rapid and accelerating pace of scientific and technological development presents policy makers, military organizations and the political system with fundamentally different phenomena and choices: the interactions of many factors over time produce an unacknowledged complexity.[7]

Bernard Brodie noted that in the span of only 22 years the US military gave up using horses to tow artillery and adopted ballistic missiles and nuclear weapons. He contrasted this experience with that of Lord Nelson:

> When Admiral Nelson was killed at Trafalgar in 1804 aboard the flagship Victory, the ship was then forty years old. Of course it had been rebuilt several times because of rotting timbers, but it was the same ship in design, and it had exactly the same guns that it carried for forty years, smoothbore 32-pounders which fired only solid, round shot. Thus, Admiral Nelson could learn his trade and exploit it without fearing that technology would take the ground out from under his feet.[8]

For years, national security analysts have warned that accelerating social change would create new and different kinds of problems.[9] The stability that comes from dealing with well-known problems and methods no longer exists for national security officials.[10] As the size of the US political-economic system has grown, the numbers, interactions, and combinations of products, technologies, and activities have multiplied dramatically. Applied to the military arena, the process of invention and incremental improvement of new technologies and capabilities across a wide range of equipment and tasks has created an interconnected, self-generating or self-reinforcing dynamic of change, where each invention or improvement leads to still others. The cycle overlaps and accelerates, providing further opportunities for novel products and tasks and the concomitant extinction of superseded tasks and products.[11] This self-generating dynamic affects military and civilian organizations alike; "as one sub organization responds in some fashion to changing conditions, its response creates new circumstances for fellow sub-organizations. As they react, new conditions affect still other parts of the encompassing organization." [12] Taken together, the processes of invention, incremental change, and organizational response act as a positive feedback, where actions are amplified beyond those originally anticipated, designed, or desired. [13]

In an effort to maintain control, both senior executives and system designers seek to constrain positive feedback, yet a self-generating dynamic of modern technological change profoundly affects the efforts of decision makers and organizations working toward qualitative improvements in military capability. Managerial efforts to impose control over innovation do not succeed. Despite heroic efforts to anticipate the pace and direction of military technology and doctrinal development (and to plan accordingly), military revolutions are most clearly identified with hindsight. [14] Indeed, philosopher of science Sir Karl R. Popper has shown that specific secondary and higher order effects of self-generating technological and social change are impossible to predict. [15]

Earlier military revolutions came about in an environment in which organizations identified and corrected doctrinal errors through a years-long process of trial-and-error and incremental change. In this way, policy makers and military professionals overcame technological uncertainties inherent in the development of new weapons. Today's national security and congressional policy makers are much like Mr. Jourdain, Moliere's protagonist in Le Bourgeois Gentilhomme, who did not know he spoke prose. In their concern for meeting defined goals, these policy makers do not recognize the self-generating dynamic of technological change that governs their progress. Very complex man-machine-organization systems inevitably beget novelty-we cannot stop it. In such systems we must manage-not control-self-generating technological change, understanding that goals and plans will be superseded and that a process of reasoned criticism will be necessary to identify errors, to highlight unanticipated implications, and to propose effective solutions.

Maintaining a distinction between management and control is crucial to fostering innovation-military or otherwise. The concepts of management and control are not synonymous. In organizations, the concept of control implies the ability to determine phenomena and events. Yet the extent to which bureaucratic control can be exercised depends upon knowledge of cause and effect relations and associated procedures to apply that knowledge. With reliable knowledge, the "manager" needs only to ensure compliance with the procedures. However, in military acquisition, complete, reliable, and verified cause-effect knowledge relating emerging or "innovative" technologies and operational concepts to the results of combat does not exist in the early stages of a project. Sometimes such knowledge exists only after the evaluation of combat outcomes. The concept of management, in contrast, assumes incomplete knowledge, uncertainty, and the consequent necessity of more flexible responses to problems. Management functions occur because important organizational problems are risky and not under control. [16]

Ironically, too often an organizational and decision process capable of such flexibility looks messy and undesirable. Senior leaders commission stand-alone studies to forecast the direction of technology and to identify those factors that can be manipulated to increase combat capability. [17] The outcomes of such studies become enshrined in acquisition programs where they act to constrain revolutionary change. [18] Absent is a continual learning process of interaction and exchange among relevant groups or agencies that evaluate goals, exercises, experiments, and simulations and channel these results back into original plans.

Policy makers seeking to manage and encourage revolutionary military advances compound the recurrent irrelevance of technology forecasts by overemphasizing management techniques of control as a means of error correction. [19] On the one hand, individuals and organizations that are intolerant of error are unlikely to identify and nurture an immature, but revolutionary complex of technologies and operational concepts. They treat unanticipated results as errors that are inconsistent with established goals-for example, doctrine or the procurement of particular technologies. Risk-loving individuals and organizations, on the other hand, may be open to unanticipated results, but are less likely to identify technological or operational failures quickly enough.

Identifying errors in the military acquisition decision process hinges on enhancing policy makers' ability to apply knowledge and analysis to problems. Experience suggests that policy analysis should be directed to improving the interaction among individuals and organizations-that is, to highlight intelligent criticism of programs, while resisting the tendency to allow a single organization (or individual) to make decisions by intellectual fiat. [20] Aaron Wildavsky framed this issue by contrasting policy analysis that "proceeds by recommending change in the structure of social interaction" with analysis that advises "larger doses of direct control by bureaucratic orders under the guidance of intellectual cogitation." [21] As applied to military acquisition, the structure of social interaction, over the long run, will permit an effective accumulation of knowledge about the efficacy of force structure mixtures composed of weapons, operational concepts, and personnel.

The Representation Problem

How easily a solution to any problem is achieved depends largely on how the problem is represented. For example, arithmetic calculations of all sorts became much easier when Arabic numerals and place notation replaced Roman numerals. [22] In the case of military innovation, traditional academic disciplinary boundaries, employed by national security analysts, have impeded the necessary analysis of interactions among variables. To understand the unfolding of military innovation and the emergence of a military revolution, American military analysts must combine a variety of perspectives. The interplay of these approaches influences how well we understand the outcomes of efforts to innovate.

• New ideas or technologies are conceived and implemented within the constraints and opportunities afforded by the American constitution and process of government. The use of knowledge and analysis in decision-making, the coordination of action among disparate groups, and the effort to lobby to wield influence are critical political processes that influence innovation.

• All decision makers involved in innovation must confront questions of evidence and knowledge. What do we know about the phenomena, for example, the chemistry and physics, in question? What are the theoretical limits to our knowledge? What are the practical limits to our knowledge? What inferences are appropriate to the data and evidence available? What evidence is appropriate to evaluate requirements, the status of the project at any time, and projections of progress?

• The strategy for overcoming uncertainty in the design and production of a new weapon or concept of operation is affected by the way an organizational structure conceives and handles errors. Explaining the process of innovation involves descriptions of how people organize to learn and evaluate their efforts over time. Specific characteristics of the "innovating" organization are important matters of investigation, including the role of hierarchy in identifying and responding to problems and the types (e.g., code, channel, calculation, or command) and distribution of administrative redundancies. [23] Particular problems of innovation include the well-known problems of goal displacement, uncertainty absorption, and coordination and coalition formation.

• Ideas, behavior, and experience of individuals over time, and accidental and random historical factors, affect the acceptance and implementation of new technology. These concerns of history are closely linked to the study of organizations—the sequence of events depends on where the system is, where the system has been, and the way in which key factors act and interact.

• Military innovation takes place within a particular political and economic regime. To explain technological change, economic historians explore variables that determine which societies become technological leaders and how long such leadership lasts. They may similarly identify variables that determine which military organizations adjust effectively to continuous technological change.

Some scholars argue that there is no grand theory of innovation. [24] Yet, the absence of a theory at present cannot imply that there will be no such theory in the future. In fact, many useful generalizations contribute to understanding military innovation. Key matters for discussion include the distinction between initiation and implementation; the role chance plays in technological evolution, time horizons, and cultural and organizational prerequisites for innovation.

Initiation and Implementation

Until the mid-1970s, researchers analyzing innovations in organizations focused upon the decisions of an individual— usually a senior leader—to adopt or reject an idea, program, or technology independent of decisions and actions of others in the organization. [25] Over the succeeding years, this simple approach to understanding and explaining innovation has been supplemented by more complex analyses. Not surprisingly, process studies of innovation have shown that, even when a decision has been made to initiate a new program, its implementation is not a certainty. Hence, analytic attention must be devoted to the people at the lower levels of organization who implement policies and programs. [26]

Other studies have detailed the relationship of various organizational structures to decisions and actions of individuals to initiate and implement innovations. [27] For example, centralization and formalization of decision-making restricts the range of new ideas considered by an organization, and thus restricts the introduction of new ways of working. However, low centralization is associated with difficulty in implementing an innovation, because the direct level of oversight or review of performance is lower. [28] In addition, an organization's members will propose more innovations to the degree to which they possess a relatively high level of knowledge and expertise (as measured by the number of occupational specialties). But, such specialization also makes it more difficult to achieve consensus about implementing those proposals. [29]

Everett Rogers proposed a simple, but useful, model of innovation that distinguishes tasks performed in initiation from those performed in implementation. Much research into innovation over the last 30 years can be placed comfortably into this model. The initiation phase in military innovation is analogous to what political scientist Nelson W. Polsby calls "incubated innovation," as efforts are made to search policy options systematically, to relate means to ends, to consider alternatives, and to test alternatives against goals. Initiation-phase tasks include agenda setting and matching a solution to a problem. Chance also plays an important role in originating new ideas and bringing these ideas to the attention of people or organizations who can further them. [30] Organizational arrangements, rules, or procedures that make it easy to reject new ideas militate against innovation. Flat organizational structures where the "occupants of the various centers of power" disagree sharply over preferred outcomes may be considerably more conducive to the research and advocacy upon which the initiations of military innovations are founded. [31]

In implementation, Rogers describes tasks including fitting the innovation to the organizational setting, clarifying its meaning, and making the innovation routine so that it is no longer new. [32] It is frequently difficult for organizations to copy successful programs or styles carried out elsewhere. A lot of management literature is devoted to creating strategies to overcome resistance to innovation. [33] An innovation must be much more beneficial than the existing procedure or technology before the "flow of benefits compensates for the relative weakness of the newer" organizational relationships mandated by the innovation. [34] The process of inventing new roles has high costs in terms of worry, time, conflict, and temporary inefficiency. [35] Where the "liability of newness" is will tend to be carried out only when the alternative to innovation is bleak. [36]

Several factors may mitigate the liabilities of newness to organizations, including (1) the capacity of individuals to learn new roles, (2) the ability to recruit people having the necessary skills, and (3) the distribution of skills in the population outside the organization. Innovations are implemented successfully when there is strong support within the organization (including among the rank and file), for the innovation and the innovation becomes part of the core practice of the organization. [37] The organization must be able to communicate and implement ideas, which then become part of the understood and accepted organizational routine. [38]

The extent and timing of oversight and review are critical to successful innovation. Detailed oversight and review occurring too early can discourage or impede the implementation of innovation. [39] Too many veto points in other parts of the organization or in society at large can stifle an innovation to rove its value. [40] Hence, isolating the innovative group often promotes success, presumably because the negotiation, transaction, and coordination costs are thereby minimized. [41] At the time, some oversight or review is necessary to avoid other types of failures: schedule delays, cost overruns, performance shortfalls, or mismatch between requirements and operational capability. [42]

Chance Factors in Technological Evolution

As Ernst Mach noted almost one hundred years ago, many critical discoveries have depended on fortuitous events that "were seen numbers of times before they were noticed. "[43] Chance events or actions affect the invention or implementation of innovations in several ways. Historical accidents sometimes enable a technology to gain an early lead over competing technologies to "corner the market" and lock potential competitors out of consideration. An established technology may become so dominant that superior alternatives developed subsequently cannot supplant it. Examples of dominant, but inferior, technologies include the narrow gauge of British railways, the 1950s programming language FORTRAN, and the QWERTY keyboard. [44] A military instance of the continued dominance of an inferior technology may be found in the way the interwar Royal Navy designed and adopted aircraft for carrier operations. Senior Royal Navy officers did not ask whether more effective aircraft designs and operational concepts existed or could be designed. [45]

Whether an organization or society will consider replacing an inferior, but familiar, technology partly depends on how easy it is to redirect the advantages associated with the inferior technology. Where those advantages involve high learning costs and specialized fixed costs (as with the QWERTY keyboard), change is difficult to implement. Where advantages from the use of an inferior technology derive from coordination—everyone uses it—change is easier as long as all can be convinced the new way offers advantages. The presence of a central authority—for example, the secretary of defense—can ease adoption of a superior technology by enforcing cooperation, but the presence of a central authority does not guarantee that inferior technologies will not sometimes prevail. [46]

Chance factors also may influence the fate of a superior technology. The mere availability of world-leading technology does not guarantee its successful exploitation if it is controlled by individuals or organizations unable to see its potential. Before World War II, for instance, American Telephone and Telegraph Company (AT&T) had a world lead in magnetic recording devices. Corporate customers and outside companies tried to interest AT&T in producing and selling such devices. Yet, senior AT&T executives suppressed the commercial exploitation of magnetic recording because they believed that the availability of a recording device would make "customers less willing to use the telephone system and so: undermine the concept of universal service." [47] These beliefs, of course, were not subjected to empirical examination or any sort of test. Nor did AT&T have an organizational structure amenable to testing, probing, or examining the premises behind such strategic decisions. The consequences of their miscalculation are obvious.

Time Horizons

In both military and nonmilitary innovation, often takes many years for a set of technologies that challenges quo to mature. [48] During the transitional period, good arguments may be presented for continued reliance upon older technology and concepts. For example, facing a shortage of black powder in 1775, Benjamin Franklin recommended the use of bow and arrows over muskets against the British. The bow presented several advantages over the musket: a good archer could be discharged in the time it took to load and discharge a musket round, smoke did not obscure an archer's view, a rain of arrows falling on an enemy had a terrifying effect, and bows and arrows would be supplied much faster than musket, ball, and powder. If the American revolutionaries had black powder from France and Netherlands to persecute the war against Britain, Franklin's proposal might have been implemented. [49]

Some major nonmilitary technological innovations have been introduced very rapidly, but these cases accentuate the difficulty of implementing military innovation, and the length of time it takes to alter an existing relationship between offense and defense. [50] In the early 1960s, for example, the National Aeronautics and Space Administration (NASA) invented, developed, and implemented much new

9

technology, including the means to launch and protect humans in space, launch earth-orbiting spacecraft and send them to distant planets, and provide supporting communications and control. [51] NASA's tasks were assigned under favorable conditions: (1) technical uncertainties were well-defined and means existed to resolve them, (2) NASA did not have to abandon or modify existing equipment and doctrine, and (3) great political certainties structured NASA's activities in the 1960s. NASA's advantages in inventing and implementing new technologies highlight the many obstacles to implementing military innovations. Military planners often are uncertain concerning what they need to know. The military has a heavy investment in older technologies and operational concepts, and political uncertainty always complicates and exacerbates the difficulty in solving technical questions. [52]

Innovations may be accepted within one or two years if the sources of ideas are people close to those agencies responsible for enactment, little time or effort is devoted to research, and every feasible alternative is not tested. [53] Post-World War II innovations in military technology, however, often have featured the coordination of people far from the ultimate decision makers, a lack of early agreement on the need and specifics for action, and a great deal of time and effort devoted to study and analysis. And so years pass between the initiation and implementation of military innovations.

Cultural Requirements for Innovation

An underlying cultural disposition favoring the application of rational thought to problems is critical to fostering innovation, as is a political system that embodies incentives to search for innovations. These two factors exist to a large extent in the United States, although conditions unique to particular situations may retard invention or implementation of innovations. Chapter 3 examines the political factor, arguing that incentives to search for innovations are incorporated into American constitutional routines associated with the electoral cycle and the separation of powers. [54]

To understand the cultural and organizational requirements for innovation, we must recognize that a powerful cluster of norms and biases make innovation possible, including the cognitive norm that "causes have effects." [55] The unconscious acceptance of such norm and biases makes it easier to reason about planning and stating objectives, creating tools, and accomplishing tasks-and thus, to manipulate the world in which we live.

A comparison can be made between the intellectual task of those charged with administering economic development programs and those responsible for military innovation. Administration is a directive process in each, that is, a causal agent of change. In both fields, the difference between effective and ineffective modes of thought and decision-making is a function of decision premises and rules of evidence. [56] Martin Landau argued that all observers of development administration establish an empirical outlook as a condition for achievement, regardless of whether they examine modernization and development from the perspective of economics, political science, or sociology. [57] Lucian Pye observed that an empirical orientation "is the essence of what we think of as modem life." [58] And Wilbert Moore added that the "institutionalization of rationality," a problem-solving orientation whereby ends and means are correlated deliberately, "is a condition for even getting started." [59]

An empirical orientation entails epistemological assumptions regarding the inculcation of a "rational" mode of analysis into one's decision calculus. Rational analysis includes an accurate description of the current situation to be acted upon, a clear statement of the desired goal or objective, and the means to eliminate or reduce the differences between the current and desired states. In development administration, a significant obstacle to creating an empirical outlook is that the introduction of rational modes of analysis disrupts and threatens to transform the most fundamental features of traditional societies. Experience has shown the great difficulty of transforming the epistemological basis of societies; the conflict between Western and Islamic values in Muslim countries is a case in point.

Similarly, exploiting new technology and the advancing rate of technological change in the United States requires the application of empirical premises and assumptions to military acquisition. [60] In practice, empiricism demands testing, hypothesis and experimentation, and trial and error as necessary and indispensable modes of behavior. Yet, as in traditional societies challenged by development projects, obstacles exist in military organizations to testing, experimentation, and self-evaluation. [61] In both cases, one may observe the ease with which people substitute rhetoric for rationality. Significant cultural shifts that promote rationality require a widespread educational effort, underlining the importance of a professional military education that imparts norms of empiricism. While education may be necessary to facilitate the promotion of rational analysis, it is not a sufficient condition.

Policy makers substitute rhetoric for rationality in acquisition decisions because it is often difficult to distinguish organizational outputs from outcomes, and evidence relating organizational structures to outcomes is extremely hard to collect. Outputs concern the work an organizational component does, for example, producing widgets. Organizational outcomes concern the results of the organizational component's actions. In general, during wartime, it is not easy for senior commanders to determine either outputs or outcomes. Outputs are hard to observe because soldiers, sailors, and airmen act out of view of the senior commander. Results or outcomes are difficult to determine because the organization usually lacks a reliable and proven method to gather information about the consequences of its actions, the outcome results from an unknown combination of operator behavior and other factors, or because the outcome appears after a long delay. [62]

James Q. Wilson proposed a typology of organizations based upon the concepts of outputs and outcomes. Four kinds of organizations exist: (1) production (both outcomes and outputs can be observed), (2) procedural (outputs but not outcomes can be observed), (3) craft (outcomes but not outputs can be observed), and (4) coping (neither outputs nor outcomes can be observed). Military organizations are generally procedural in nature, where managers can observe their subordinates' actions, but not the outcome from the efforts. In Wilson's words,

> Perhaps the largest procedural organization in the government is the United States Armed Forces during peacetime. Every detail of training, equipment, and deployment is under the direct inspection of company commanders, ship captains and squadron leaders. But none of these factors can be tested in the only way that counts, against a real enemy, except in wartime. [53]

In wartime, military units change from procedural to craft organizations consisting of operators whose activities are difficult to observe but whose outcomes are relatively easy to evaluate. [64] But, in peacetime, a military service cannot "afford to allow its operators to exercise discretion when the outcome of that exercise is in doubt or likely to be controversial." Because management is constraint-driven, senior leaders become means-oriented. Thus, "how the operators do their jobs is more important than whether doing those jobs produced the desired outcomes." Standard operating procedures (SOP) become pervasive. Wilson also noted "in recent decades the US Army has devoted much of its peacetime efforts to elevating SOPs to the level of grand tactics by trying and then discarding various war-fighting doctrines. But when war breaks out, SOPs break down. The reason is obvious: outcomes suddenly become visible." [65]

Successful innovation also requires the presence of groups capable of inventing, developing, and marketing a new product or activity. [66] Bureaucratic entrepreneurs often find it necessary to devise the symbolic, organizational, or procedural means to protect an innovative group and its new ideas from criticism and review-developing an esprit de corps or a sense of community may be necessary for an innovation to take hold. [67] A sense of community was an essential ingredient to the success of Adm Hyman Rickover's efforts to build a nuclear-powered Navy. This success was obtained partly because membership in the community created barriers to observation, evaluation, and interference in programmatic activities by "outsiders"—people not in the community. A sense of community also encouraged the formation of trust, so that honest internal programmatic evaluation could take place.

It is interesting to speculate whether Rickover's success eventually altered the prospects for others attempting military innovation. Frustrated at dealing with Rickover, political and Navy officials may have modified institutional rules, making it less likely that such independent leadership would be exercised in the future. Indeed, institutional rules may permit several organizational styles; Rickover's tenure overlapped with the Navy's Special Projects Office which operated differently and did not have the same bureaucratic effect on " the Navy or on the other services.

Summary

The ultimate fate of an emerging military revolution is tied, to the performance of the economy, society, and political system. As the twentieth century closes, the acceleration of knowledge and technology has created more opportunities for invention and implementation of innovation. Yet, the management of innovation can be hobbled by chance or accidental factors, short time horizons, bureaucratic vetoes, and poor decision-making. Innovations embody both specific technical knowledge and a particular social, economic, and political context. Even in the presence of technical knowledge or insight, the absence of an appropriate political and social setting has retarded acceptance and implementation of many inventions. Many weapons or tools have waited years for the emergence of an appropriate social structure to make their employment possible. [68]

National security decision makers face complex and risky choices. The complexity of these choices underlines the importance of leadership and the factors within an organizational structure that either encourage or limit thinking and deciding. Some analysts call on national security decision makers to look beyond current problems and to propose options and solutions to problems that do not yet exist. [69] Such proposals make unreasonable demands upon the cognitive capacity of decision makers, as decisions shaping the future result from solutions to immediate problems. Exhortations to plan a synoptic or comprehensive reassessment of technology, management, and organizational opportunities often lead to rhetoric instead of rationality.

Innovation is not the result of incantations and magic. It is a process that can be understood. This chapter reviews some issues basic to understanding the character of military innovation in the late twentieth century. The next chapter argues that attention to distinct levels of analysis or units of analysis (individual, organization, and institution) and the interaction among them are also critical to understand innovation.

Notes

1. Webster's New Twentieth Century Dictionary of the English Language, unabridged, 2d ed. (Cleveland: World Publishing Co., 1971), 945.

2. C. T. Onions, ed., The Oxford Dictionary of English Etymology (New York: Oxford University Press, 1966), 476.

3. Henry Petroski, The Evolution of Useful Things (New York: Alfred A. Knopf, 1992).

4. Eric von Hippel, "The Dominant Role of Users in the Scientific Instrument Innovation Process," Research Policy, 1976, 212-39.

5. James G. March and Herbert A. Simon, Organizations (New York: John Wiley & Sons, Inc., 1958), 174-75.

6. Military technology improved incrementally. Artillery could be maneuvered more easily on the battlefield and infantry firearms became more reliable and accurate. Russell F. Weigley, The Age of Battles: The Quest for Decisive Warfare from Breitenfeld to Waterloo (Bloomington: Indiana University Press, 1993).

7. Eugene B. Skolnikoff, The Elusive Transformation: Science, Technology, and the Evolution of International Politics (Princeton, N.J.: Princeton University Press, 1993).

8. Bernard Brodie, "Introductory Remarks," in Science, Technology, and Warfare, ed. Monte D. Wright and Lawrence J. Paszek (Washington, D.C.: Government Printing Office [GPO], 1970), 85.

9. For example, Quincy Wright, A Study of War, vol. 1 (Chicago: University of Chicago Press, 1942), 4.

10. As Wilbert E. Moore notes, " in the modem world social change has taken on some special qualities and magnitudes." Social Change (Englewood Cliffs: Prentice-Hall, Inc., 1963), 2.

11. Stuart A. Kauffman, "The Evolution of Economic Webs," in The Evolution as an Evolving Complex System, ed. Philip W. Anderson, Kenneth J. Arrow, and David Pines (Redwood City, Calif.: Addison-Wesley Publishing Co., Inc., 1988), 126, 139-42.

12. Wallace S. Sayre and Herbert Kaufman, Governing New York City: Politics in the Metropolis (New York: Norton, 1965), xlv-xlvii: Herbert Kaufman, "Chance and Organizational Survival: An Open Question (Part II)," Journal of Public Administration Research and Theory I, no. 3 (July 1991): 367-68.

13. Economists might call this phenomenon an example of "increasing returns on margin." Other synonyms include self-reinforcement, deviation-amplifying mutual causal processes, and cumulative causation. W. Brian Arthur, "Self-Reinforcing Mechanisms in Economics," in The Economy as an Evolving System, ed. Philip W. Anderson, Kenneth J. Arrow, and David Pines (Redwood City: Addison-Wesley Publishing Co., 1988), 10. Negative feedbacks are error corrective and are defined in terms of a goal, measurement device (to determine the difference between the present state and goal), and effectors (to nudge the system toward its goal). The importance of negative feedback cannot be overstated, and a great deal of effort (e.g., in philosophy, cybernetics, electronics, and physiology) has been devoted to its analysis.

14. Norman Friedman, Thomas C. Hone, Mark D. Mandeles, The Introduction of Carrier Aviation into the U.S. Navy and Royal Navy: Military-Technical Revolutions, Organizations, and the Problem of Decision (Washington, D.C.: Office of the Secretary of Defense [OSD]/NA, 1994), 4-5.

15. Karl R. Popper, The Poverty of Historicism (New York: Harper Torchbooks, 1964).

16. Martin Landau and Russell Stout Jr., "To Manage is not to Control: Or the Folly of Type II Errors," Public Administration Review 39, no. 2 (March/April 1979): 148-56.

17. See, for example, Robert Holzer, "U.S. Navy Pursues Technology Leaps," Defense News, 25-31 July 1994,14.

18. Agencies like the US General Accounting Office (GAO) abet this constraint to revolutionary change by conducting deficiency audits-audits that focus solely upon the difference between intended and actual outcomes.

19. For example, Landau and Stout; Chris C. Demchak, Military Organizations, Complex Machines: Modernization in the U.S. Armed Services (Ithaca: Cornell University Press, 1991).

20. Taking a cue from recent chaos research, complex systems are more readily influenced if changes are addressed to their margins. See the discussion of how chaotic systems may be controlled in William L. Ditto and Louis M. Pecora, "Mastering Chaos," Scientific American 269, no. 2 (August 1993): 78-84.

21. Aaron Wildavsky, Speaking Truth to Power: The Art and Craft of Policy Analysts (Boston: Little, Brown & Co., 1979), 113.

22. Herbert A. Simon, The Sciences of the Artificial, 2d ed. (Cambridge: The M.I.T. Press, 1990), 150-53; John A. Paulos, Beyond Numeracy (New York: Vintage Books, 1992), 15-17.

23. In like manner, noted economist Richard R Nelson argues that "intrafirm organization" is an important variable "in its own right" in explaining research and development (R&D) and productivity growth. See Production Sets, Technological Knowledge, and R&D: Fragile and Overworked Constructs for Analysis of Productivity Growth," American Economic Review, May 1980, 62-67. Martin Landau effectively introduced the concept of redundancy and applied a typology of redundancy (originally proposed by physiologist Warren McCulloch) to the study of organizations. See Landau, "Redundancy, Rationality, and the Problem of Duplication and Overlap," Public Administration Review 29, no. 4 (August 1969): 346-58; Landau, "Linkage, Coding, and Intermediacy: A Strategy for Institution Building," Journal of Comparative Administration, February 1971, 401-29.

24. Stephen P. Rosen, Winning the Next War: Innovation and the Modern Military (Ithaca, N.Y.: Cornell University Press, 1991), 1-5; A. W. Coats and David C. Colander, "An Introduction to the Spread of Economic Ideas." ed. David C. Colander and A. W. Coats (Cambridge: Cambridge University Press, 1987).

25. Everett M. Rogers, Diffusion of Innovations, 3d ed. (New York: Free Press, 1983), 355.

26. For example, Pressman and Wildavsky showed that even though a large number of individuals in different agencies favored adoption of a particular program, its implementation was no easy matter. Jeffrey L. Pressman and Aaron B. Wildavsky, Implementation (Berkeley: University of California Press, 1973); see also James Q. Wilson. Bureaucracy: What Government Agencies Do and Why They Do It (New York: Basic Books, Inc., 1989), 11.

27. Rogers, 348.

28. Ibid., 359-61.

29. Ibid., 360.

30. Coats and Colander; Dean Keith Simonton, "Creativity, Leadership, and Chance," in The Nature of Creativity: Contemporary Psychological Perspectives, ed. Robert J. Sternberg (Cambridge: Cambridge University Press, 1988), and 387.

31. Nelson W. Polsby, Political Innovation in America: The Politics of Policy Initiation (New Haven: Yale University Press, 1984), 155.

32. Rogers. 362-65.

33. Kaufman, "Chance and Organizational Survival," 361.

34. Sir Charles Carter, in his presidential address to the Association for the Advancement of Science, argued that the economic system has been unable to mitigate the liability of newness. He noted that the British too frequently attempt a too great leap in technology-before the benefits of the new way of doing things become evident. "Conditions for the Successful Use of Science." Science, 18 March 1983. 1296. See also Arthur L. Stinchcombe, "Social Structure and Organizations." in Handbook of Organizations, ed. James G. March (Chicago: Rand McNally & Co. 1965), 148.

35. Stinchcombe, 148.

36. Ibid, But, March and Shapira note that there is little literature support for the proposition that organizations accept riskier alternatives when in trouble, or that they innovate when in trouble, James G. March and Zur Shapira, "Behavioral Decision Theory and Organizational Decision Theory." in Decision Making: An Interdisciplinary Inquiry, ed. Gerardo R. Ungson and Daniel N. Braunstein (Boston: Kent Publishing Co., 1982), 108.

37. Daniel A. Mazmanian and Paul A. Sabatier, Implementation and Public Policy (Glenview, Ill.: Scott Foresman & Co., 1983), 277; Robert K. Yin, "Life Histories of Innovations: How New Practices Become Routinized," Public Administration Review, January/February 1981, 26.

38. Simonton, 396.

39. James Brian Quinn, "Technological Innovation Entrepreneurship and Strategy," Sloan Management Review, Spring 1979, 22.

40. Mazmanian and Sabatier, 266. In this context. Rosen cites a memorandum written by Secretary of Defense Robert S. McNamara to Army Secretary Elvis J. Stahr Jr. concerning a study group to consider the role of helicopters in ground operations. McNamara wrote, "it requires that bold, new ideas which the task force may recommend be protected from veto or dilution by conservative staff review." McNamara clearly was worried about the type of review that could reject a good idea before the idea had time enough to prove itself. Rosen, 86.

41. Harvey Sapolsky, "Organization Structure and Innovation," Journal of Business, 1967, 497-510.

42. US GAO. "Defense Acquisition: Oversight of Special Access Programs Has Increased." GAO/NSIAD-93-78, December 1992. See also "Special Access Programs: DOD Criteria and Procedures for Creating Them Need Improvement." GAO/NSIAD-88-152, May 1988; "Special Access Programs: DOD is Strengthening Compliance With Oversight Requirements," GAO/NSIAD-89-133, May 1989.

43. Emphasis in the original. Ernst Mach, Monist (1896), 169, cited in Simonton, "Creativity, Leadership and Chance," 396.

44. W. Brian Arthur, "Competing Technologies, Increasing Returns and Lock-In By Historical Events," The Economic Journal, March 1989, 126; and Paul A. David, "Clio and the Economics of QWERTY," American Economics Review 75, no. 2 (May 1985): 332-37.

45. Friedman, Hone, and Mandeles, chap. 7.

46. Arthur, "Self-Reinforcing Mechanisms in Economics," 26.

47. Mark Clark, "Suppressing Innovation: Bell Laboratories and Magnetic Recording," Technology and Culture 34, no. 3 (July 1993): 516-38.

48. For instance, it took several hundred years of improvements and evolution for today's simple pencil to be able to be manufactured. Heruy Petroski, The Pencil (New York: Alfred A. Knopf, 1989); see also Petroski, The Evolution of Useful Things.

49. Norman B. Wilkinson, Explosives in History (Chicago: Rand McNally & Co., 1966), 14.

50. Clifford J. Rogers, "The Military Revolutions of the Hundred Years War," Journal of Military History 57, no. 2 (April 1993): 241-78.

51. Leonard R. Sayles and Margaret K. Chandler, Managing Large Systems: Organizations for the Future (New Brunswick: Transaction Publishers, 1993), xviii.

52. Howard E. McCurdy, Inside NASA: High Technology and Organizational Change in the U.S. Space Program (Baltimore: Johns Hopkins University Press, 1993).

53. Polsby, 150-53.

54. Ibid., 165.

55. Ibid., 160.

56. The epistemological or methodological assumptions of traditional and modem societies have been the topic for many scholars. See Martin Landau, "Development Administration and Decision Theory," in Development Administration in Asia, ed. Edward Weidner (Durham: Duke University Press, 1970), 72.

57. Landau, "Development Administration and Decision Theory," 82.

58. Lucian W. Pye, "Introduction," in Communications and Political Development, ed. Lucian W. Pye (Princeton, N.J.: Princeton University Press, 1963), 19.

59. Moore, 95.

60. This recommendation is consistent with Jacob A. Stockfisch's arguments about defense analysis. For example, see "Operational Testing," Military Review, May 1971, 66-82; "The Intellectual Foundations of Systems Analysis," P-7401 (Santa Monica: The RAND Corporation, December 1987).

61. Aaron Wildavsky, "The Self-Evaluating Organization," Public Administration Review 32, no. 5 (September/October 1972): 509-20; reprinted in Aaron Wildavsky, Speaking Truth to Power (Boston: Little, Brown & Co., 1979).

62. Wilson, 158-59.

63. Ibid., 163.

64. Ibid., 165.

65. Ibid., 164.

66. The presence of such groups, representing enclaves of innovation, seems to have been necessary to establish the economic and social viability of empires. Such elites or "entrepreneurs" aided the rulers of societies by generating increased resources (wealth) for the use by those rulers and society. S. N. Eisenstadt, The Political Systems of Empires (New Brunswick: Transaction Publishers, 1993), chap. 12, "Processes of Change in the Political Systems," 312-16, 359-61.

67. Harvey Sapolsky, The Polaris System Development: Bureaucratic and Programmatic Success in Government (Cambridge: Harvard University Press, 1972); Eugene Lewis, Public Entrepreneurship: Toward a Theory of Bureaucratic Political Power (Bloomington: Indiana University Press, 1980); Sayles and Chandler, 318; Gerald Garvey, Facing the Bureaucracy: Living and Dying in a Public Agency (San Francisco: Jossey-Bass, 1993).

68. Stinchcombe, 160; Oliver E. Williamson, "Emergence of the Visible Hand: Implications for Industrial Organization," in Managerial Hierarchies, ed. Alfred D. Chandler and Herman Daems (Cambridge: Harvard University Press, 1980), 189; Vincent Foley, George Palmer, and Werner Soedel, "The Crossbow," Scientific American, January 1985, 104.

69. Paul Bracken, "The Military After Next," The Washington Quarterly 16, no. 4 (Autumn 1993): 157-74. Such proposals represent a "historicist" approach to knowledge. See Popper.

Chapter 3
Logic and Procedure of Analysis

Historians and social scientists typically pose alternate interpretations of military innovations. Debates between the two groups frequently are characterized as disagreements about the truth or falsity of some interpretation of events. In reality, many of the exchanges of criticisms are really disagreements over levels of analysis. Confusion over levels of analysis has led some analysts to believe they have refuted some claims when they have not. [1] In particular, historical studies of military innovation generally focus on the activities and experiences of individuals. Frequently, the extent to which constraints and opportunities in organizations and institutions influence individual actions and decision-making is obscured in historical studies of innovation. Yet at each level of analysis entirely new properties appear. [2] A more accurate description of the innovation process would account for the separate roles of institutions, organizations, and individuals.

Nobel laureate Douglass C. North defined institutions as society's "rules of the game." Institutions reduce uncertainty by establishing a stable, but not unchanging, structure to human interaction, encompassing contracts between individuals, codes of conduct, conventions, norms of behavior, common law, statute law, and constitutional law. These rules set up the incentives and disincentives that create opportunities or constrain action for individuals and organizations. Changes in institutions affect the way societies evolve. [3] The continuity of society's institutions connects present and future choices to the past. History is important, not only because one can learn from examples and analogies drawn from past experiences but also because an examination of the past reveals the initial conditions that structure present and future choices.

The study of institutions provides a standpoint from which one may address various questions ranging from the economic performance of whole societies to the factors that mitigate or enhance entrepreneurial activity, and thus innovation. [4] Reference to institutions thus becomes critical in describing and explaining how military revolutions come about, because an institution's rules or laws affect how individual military entrepreneurs and people in organizations perceive opportunities and obstacles to coordinating action. Analysis of innovation and technological change, therefore, should examine the role of institutions in (1) influencing how some technologies win out over others and (2) accepting and implementing new hardware or concepts of organization.

If institutions are the "rules," then organizations constitute the "players." Organizations are groups of individuals bound together by some purpose, for example, political (parties and government agencies), economic (firms and trade unions), social (clubs and churches), and educational (schools and universities). Where institutions determine society's opportunities, organizations arise to exploit those opportunities. Thus, institutions influence which organizations come into existence and how they evolve. In turn, the interaction of extant organizations influences the evolution of institutions.

We must pay attention to the institutional, organizational, and individual levels of analysis in describing and characterizing innovations. This approach helps to avoid single-factor explanations that focus on only one of the three levels. A multilevel perspective also enables us to examine linkages among individuals' actions, the organizational contexts in which individuals are constrained or encouraged, and the incentives created by institutions that shape individual and organizational choices.

The current home video market offers an example of the difference between organizations and institutions, players and rules. [5] In Game Over, David Sheff describes how Nintendo created and seized control over large markets. Nintendo and other companies are the organizational players in a business and financial system. The rules of the Japanese business system are the institution. These rules create opportunities or constrain freedom of action. In this particular case, Japanese institutions-that is, business and financial rules-tolerate monopoly and do not impose demands for short-term profitability. [6] The contrast with US business and financial rules and laws is stark. Nintendo could not have succeeded as an American company because of the threat of antitrust action and a much greater demand for short-term profit. Indeed, some Writers argue that a major factor in IBM's underestimation of the personal computer (PC) business was hesitancy engendered by its lengthy antitrust battle with the Department of Justice beginning in 1969. [7]

The formal and informal rules that comprise institutions have mixed social and economic effects. Some rules induce productivity increases, while others reduce productivity or cause other unfavorable outcomes. For example, American nineteenth century economic history is a story of economic growth because the "underlying institutional framework persistently reinforced incentives for organizations to engage in productive activity however admixed with some adverse consequences." [8] The American constitutional (or institutional) rules that divide political authority (separation of powers and federalism) and constrain the exercise of power (checks and balances) allow a varied interaction of political and economic authority. These rules are significant in weapons acquisition, creating many players in the acquisition process, permitting organizations to influence the allocation of services and goods provided by government, and structuring access to decisions and decision makers. [9]

Analysis of Technological Competition

Technological competition may be examined usefully through the lens of multiple analytic levels. The agent of technological, economic, social, or political change is the individual entrepreneur, whose behavior is a response to the incentives embodied in institutions. The motivation for innovation is in the incremental elimination of defects and the effort to find a better way. More formally, economists speak of the impetus to innovation in terms of shifting preferences in communities (or more broadly, whole societies), shifts in the relative prices of goods or services, or expanding technological capacity. [10] For example, following World War I, a small group of naval officers was motivated to search for a better way to employ aircraft. This search catalyzed the technological competition between battleships and aircraft, and led to a growing preference among some Navy officers to use aircraft, instead of battleships, to apply firepower against enemy naval forces. The associated propensity to investigate new operational concepts was reinforced by the development of increasingly more powerful and efficient aircraft engines and better bombsights, which made aircraft a more effective platform for long-distance strikes.

Viewed from the perspective of multiple levels of analysis, technological competition is indirect. There may be little functional difference in outcome between competing equipment. At the outset of World War II, for example, both Great Britain and the United States produced a series of naval and army aircraft engines of about equal overall technical merit and military utility. [11] Yet, while performance of individual pieces of equipment may have been roughly equivalent, the US Navy-as an organization-was far more capable of resolving technological uncertainties and integrating acquisition of new types of aircraft with doctrinal strategies. [12] Likewise, in 1940, technological competition between Germany and France was not in the operational performance of military equipment; it was in organization and operational concepts. The qualitatively comparable and larger French force was defeated quickly by the operationally and organizationally superior Germans.

The historian's focus on the individual obscures an important consideration about innovation; military competition actually occurs among the organizations that develop, produce, or use the competing weapons technologies. An organization's choice of a technology may reflect differing organizational abilities-the tacit knowledge and experience of senior leaders or the capacity of the organizational structure to experiment and invest in productive knowledge-as much as the technical performance of competing equipment. [13] For example, Rear Adm William A. Moffett's tacit political knowledge was essential in gaining the cooperation and support of key persons in the Congress and the Navy to create the Bureau of Aeronautics and, later, to buffer that new organization from political attack. [14] Moffett's tacit knowledge also was instrumental in setting up a research program to build better aircraft. IS In Great Britain, Japan, and the United States, the ultimate choice of a configured set of naval aviation technologies (e.g., carrier design, aircraft engines, structural design, and payload) reflected how officials, possessing different leadership skills, championed aircraft and presented themselves in the larger political system. The story is, only secondarily, about the operational capability of the technologies themselves.

How organizations are structured to learn affects their ability to compete effectively in the innovation process. The most successful organizations often pursue multiple and I parallel approaches to a development problem, thus reducing uncertainty by identifying and eliminating poor options. An extensive empirical literature supports this argument. [16] Within a context of institutional rules that stresses empiricist habits of mind and the role of evidence, such organizations foster efficient feedback and successful comparison of multiple options.

An organization may embark and continue on an unproductive path in identifying and exploiting new technologies when institutions create disincentives to productive change and reinforce interest groups that have a stake in the status quo. [17] By analogy, systematic improvements in combat effectiveness may require an institutional context that allows for examination of many choices and the development of feedback mechanisms (e.g., war games and simulations) to identify and eliminate poor choices. The institutional context also must provide acknowledgment and rewards, allowing officers and enlisted personnel engaged in such work paths to higher status, responsibility, and authority.

Application of the Analytic Framework

Examining history from the standpoint of multiple levels of analysis permits a more rigorous approach to the question of how and where errors enter the decision process, how and where errors are identified and corrected, and how technical and policy uncertainties affect the identification and correction of errors. Two brief stories help to illustrate the importance of multiple levels of analysis in looking at innovation. The first describes Army tank development and the second recounts and compares how the Navy approached the development of carrier-based aviation.

In At Ease: Stories I Tell to Friends, Dwight D, Eisenhower recounted post-World War I experiments, conducted with George S. Patton at Camp Meade, Maryland, on tank concepts of operation. [18] Eisenhower and Patton disagreed with the approved tank doctrine, which saw the tank as a front-line infantry weapon moving at the speed of the foot soldier-about three miles per hour. Instead, they wanted fast tanks that could attack by surprise in mass, break through enemy defensive positions, and surround and take those positions from the rear, to attract military audiences to their point of view, Patton and Eisenhower began to prepare articles for publication, but their plans were not received well at Army headquarters. Eisenhower was called before Maj Gen Charles S. Farnsworth, the chief of infantry, and in his words, "was told that my ideas were not only wrong but dangerous and that henceforth I would keep them to myself. Particularly, I was not to publish anything incompatible with solid infantry doctrine. If I did I would be hauled before a courts-martial. George [Patton], I think, was given the same message." [19]

The explanation for the Army's resistance to a new and different tank concept of operation is more complicated than Eisenhower allowed. Senior Army leaders had been apprised of conflicting evaluations of tank doctrine by none other than Patton himself. Patton was the first member of the American Tank Corps during World War I. After reviewing the combat experience on 30 November 1917 at Cambrai, France, where the British had massed 324 tanks and achieved a breakthrough of German lines, Patton wrote his wife, "I feel sure that tanks in some form will playa part in all future wars," [20] Patton attacked the problem of developing tanks with characteristic zeal: investigating the technology, writing articles, and delivering lectures. [21] Yet, Patton's views about tanks were internally contradictory. He was

> so outspoken on the philosophy that tanks exist to support infantry (a view which supported then-current doctrine] that ranking generals such as (John J.)Pershing believed him, and so did important congressmen. The National Defense Act of 1920.... abolished the Tank Corps and assigned all tank units and personnel to the Infantry Branch. This dissolution took place despite the simultaneous creation of new branches such as the Air Service the Chemical service, and the Finance department. [22]

The standard interpretation of the Army's failure to develop the tank is hobbled by its attention to a single level of analysis-the individual. Understanding the organizational level of analysis, however, helps explain how and why the Army's development of tanks followed, rather than led, other countries. Some historians have noted that the question of whether tank functions should be classed with artillery, infantry, or cavalry was debated throughout the interwar period. In what is a widely accepted interpretation of events, Martin Blumenson argued that a consequence of the debate was that few advances in tank warfare were made by the US Army. The Army did not supply either sufficient numbers or quality of men and machines to permit experimentation. [23] This explanation is reasonable, as far as it goes. It implies that the accepted, but failed, option could not have prevailed on its merits; therefore something was wrong with the people or the way they individually made decisions. [24]

The standard view provides insufficient explanation for the decisions about tanks, however, and little understanding for future decision makers. The Army and its leaders were able to improve existing technologies. During the interwar period, for example, the Army developed excellent weapons based on existing designs, such as the 60 mm and 81 mm mortars, the 105 mm howitzer, and the M-l Garand rifle. [25] Yet, the Army's problem was in the way it organized itself to explore new technologies and new operational concepts. In 1927, after observing British armored maneuvers, Secretary of War Dwight Davis ordered the Army General Staff to create a mobile armored unit, but this order inaugurated only a short period of experimentation that ended in 1931. While a few enthusiastic officers (e.g., Adna R. Chaffee and Sereno Brett) recognized the potential of armor despite the technological limitations of tanks, the Army's experimentation was not used to probe weaknesses in doctrine, reevaluate roles and missions (of infantry, cavalry, or artillery), or to guide acquisition. The Army was not organized to learn how to recognize the potential of technologies, weapons, and concepts of operation which did not yet exist.

At the organizational level of analysis, it becomes clear that the Army was unable to develop and exploit the infant tank technology and doctrine because it was not organized to devise and learn about new ways of accomplishing its tasks. Furthermore, senior leaders saw no need to alter their decision process to enhance experimentation, debate, and interaction about new doctrine and technology. With institutional rules and an organizational structure that encouraged learning, senior Army leaders might have made better decisions about tank development.

The limitation of the single-level explanation and description of the Army's rejection of innovative mobile armored warfare concepts is highlighted by a comparison with the Navy's ability to overcome technological and organizational uncertainties in developing carrier aviation. By 1922, despite public posturing over the relative value of battleships and aviation, Adm William S. Sims instituted a process whereby the potential of naval aviation deployed with the fleet could be established systematically and rigorously through tactical and strategic simulations at the Naval War College. These simulations addressed various issues concerning the effective employment of aviation, including how aviation should be based and supported and how aviation might be used-given anticipated technological developments. [26] Indeed by 1923, the Navy had created and put into place a multiorganizational analytic system that encompassed a variety of organizations, including the General Board, the Fleet, the Naval War College, and the Bureau of Aeronautics. The system coordinated several organizational procedures (simulations, war games, and fleet maneuvers) to investigate and examine new proposals, and effectively conducted systematic analysis of the potential of new technology and operational concepts.

In effect, the concerns, deliberations, and questions posed by the Navy's General Board created informal rules of analysis and evidence. These senior officers constantly sought to gather information about matters they needed to decide. The efficacy of these rules was abetted by interaction and competition among the Navy's different communities. A remarkable feature of this interwar history is that the US Navy embarked on a self-evaluating course of inquiry regarding aviation despite the risk that it would be harmful to the Navy budget by leading to significant reductions in budgetary commitments to battleships. Outside organizations, including Congress, made explicit the aircraft-battleship trade-off. In modern times, some commentators argue that only an outside agency (e.g., one committed to operational testing) can engage in the type of inquiry carried out within the Navy in the interwar period. The Navy's interwar experience evaluating carrier aviation shows that an outside agency is not necessary to ensure that rigorous analysis is conducted on technological and operational possibilities.

The Navy and Army operated under identical formal institutional rules-the checks and balances and separation of powers embodied by the US Constitution. Yet, Navy organizations were able to exploit these formal institutional rules much more effectively to identify and reduce critical uncertainties, while the Army was unable to identify potential new weapons and develop appropriate operating concepts for those new weapons. The chief difference between the ability of the Navy and the Army to develop the new military technology is in the presence and use of institutional rules and organizations capable of evaluating uncertainty. The Army had no analogous institutions, rules, or organizations to the Navy's General Board, nor did the Army leaders set up a form of interaction comparable to that which existed among the General Board, the Fleet, the Bureau of Aeronautics, and the Naval War College. [27] As a consequence, the evaluation of claims, arguments, or ideas about tanks or aircraft depended more heavily upon chance encounters and conversations. Senior naval leaders, in creating the Bureau of Aeronautics and in initiating carrier simulations at the war college, did not understand that the Fleet, the Naval War College, Bureau of Aeronautics, and General Board would interact as effectively as they did. Post-World War II reorganizations suggest that US naval leaders did not recognize the usefulness of their interwar organizational relationships. [28]

Organizational and Institutional Interaction: Toward a Framework for Analysis

Little has been done to devise an analytic framework that would adequately describe the evolution of institutions that induce innovative successes as well as those that lead to economic or military stagnation and decline. Some useful work toward such a framework can be found in a theory of organizational behavior known as the "garbage can model." This model can contribute to an analysis of institutions that either constrain or provide opportunities for innovation by linking the historian's focus on individuals with the social scientist's interest in institutions (in Douglass North's sense), organizations, hierarchies, and decision making. [29]

The garbage can approach argues that various organizational processes (including the conception and representation of problems, the design of solutions, and the opportunities for people or organizations to participate in decision making) are separate streams of activity brought together by the need to make a decision. [30] The separate streams of activity and what the organization does are strongly influenced by the ubiquity of ambiguous evidence; fluid participation by officials, and problematic preferences. [31] Ambiguous evidence refers to "an organization's ignorance as to the options that are available to its decision makers and the linkages between these alternatives and their likely consequences." [32] Other sources of this type of ambiguity include unavailable, unreliable, or deceptive information and information overload. Fluid participation by officials (or staff or others) means that they will devote varying amounts of energy and attention to a given issue. They have discretion over whether (and how) to become involved in choices and decisions. Their participation is constrained by other demands on their time and attention, so that no single official dominates the decision process in all its phases, nor are all issues considered simultaneously. [33] Finally, the theory notes that bureaucracies operate with all sorts of ill-defined or inconsistent preferences. Such inconsistent and poorly defined preferences are often concealed until the need to take action forces those holding them to speak up. [34]

The metaphor of a decision or "choice opportunity" as a garbage can stresses that different mixtures of participants, problems, and solutions will come together. Moreover, once made, a decision may simply begin yet another garbage can cycle. The participants, problems, solutions, and choice opportunities within such a cycle may be somewhat stable when separate, but there may be no unique, ordered way in which they are always linked to each other. [35] The decisions made depend upon the connections among the different streams of activity. Problems, solutions, and decision makers are often connected more by simultaneity than by abstract relations between a theoretical problem and its solution. [36] In other words, timing drives decision-making.

One implication of this perspective is that the coherence of a senior leader's explicit intentions often is lost in the movement of people, problems, and solutions within an organization. [37] Since a decision outcome hinges on the mixture of "garbage" or streams of activity-the people, problems, situation, and solutions-at any particular time in the "can," the process leading to the decision has some features of an organized anarchy. Yet, the garbage can decision process is not random-it is highly structured by access (as defined by institutional rules) and situation. Some problems may take precedence over others, and the participants in a particular decision may vary over time and with opportunity. In addition, unobtrusive, quiet action may have a major effect on outcomes. [38]

The connection of a problem to a solution chosen from a menu of solutions by decision makers-called "coupling"-is key to understanding the decision process. The simultaneous flow of the four streams of activity in the garbage can explains why Stephen Rosen did not discover a clear link between a specific enemy "threat" and US weapons research and development between the 1930s and 1950s. [39] The peacetime R&D process was neither centrally directed nor focused on one well-defined problem.

The way people connect, solutions and problems also illuminates the relationship between innovation and the extent to which an organization is open to the exchange of personnel and ideas from other parts of the organization or from other organizations. Students of American government have long understood how interest groups faced with a political defeat often seek to alter the audience, enlarge the number of participants in the process, or change the arena in which the decision is being made. [40] This phenomenon also applies to bureaucratic struggles over innovation within or between agencies. For example, during the interwar years, bureaucratic battles over Brig Gen William "Billy" Mitchell's proposal to create a unified and independent air force, as part of the struggle between Army and Navy airmen, moved from negotiations between the secretaries of war and the navy to congressional meeting rooms to newspaper headlines to Mitchell's court-martial. [41]

In the process of innovation, rational, irrational, and nonrational behavior will be mixed, so that an innovation rarely will move in quite the direction or speed that its initiators anticipated. The model assumes that change is a messy, often confused process, even within structured bureaucracies having well-articulated standard operating procedures and composed of well-trained professionals. Thus, the garbage can perspective does not establish an "equation" of military innovation, where the dependent variable (the development of a set of technologies and doctrine, etc., leading to a military revolution) is a precise function of differently weighted independent variables, such as organizational hierarchy, information flow within the organization, or the nature and degree of congressional involvement. Instead, the garbage can perspective tells us to look for the interactions- to be sensitive to streams of activity that may converge for reasons impossible to predict ahead of time. In the garbage can, the different levels (individual, organizational, and institutional) may interact in unexpected and nonlinear ways.

The garbage can perspective can be generalized: it applies to societies other than in the United States. For example, although the American, British, and Japanese navies all developed carrier aviation before World War II, each was not necessarily driven by the same technological or strategic imperatives. Senior military leaders possessed different levels of tacit political skills, and applied those skills to disparate conceptions of their prime tasks. Their naval bureaucracies did not operate in the same basic way at the organizational level. Nor were their political institutions and societies essentially alike at the institutional level. But the garbage can perspective assumes that there were some similarities concerning how individuals, organizations, and institutions interact, just as there were important differences. And, in fact, the British and Japanese decision makers also faced ambiguous technology, had problematic preferences, and devoted varying amounts of time and concentration to naval aviation issues.

This perspective is crucial to developing an analytic framework for innovation because it counteracts the tendency to fix on a single level of analysis. Attention to levels of analysis permits more effective recommendations concerning organizational design to encourage innovation. Key to these recommendations may be the realization that Landau's work on self-correcting organizations is consistent with Wildavsky's on self-evaluating organizations. [42] When viewed from only the organizational level, the two seem contradictory. but their consistency becomes readily apparent when viewed from the interaction of organizations and institutional rules. Wildavsky argued that the conditions necessary for a self-evaluating organization-one capable of self-correction-cannot occur because the needs of the organization and the people within it conflict with the mandate to monitor activities continuously in an intellectually honest way and to change policies when they are ineffective. Landau, however, contended that organizations will perform best if set on the foundation of rational criticism. The seeming contradiction in these positions disappears if one considers the interaction of organizational and institutional levels. Institutional rules (e.g., professional standards of conduct, rules of evidence and inference, and a political system that tolerates or encourages exchange among organizations and individuals) influence, constrain, and guide relationships in a multiorganizational system. These rules encourage a system of organizations to examine plans, programs, and policies critically, regardless of whether the individual organizations making up the system exhibit self-correcting or self-evaluating behavior.

Thus in the carrier aviation example referred to above, no one of the American naval agencies acting on its own could have pursued carrier development to the detriment of battleships. But the system of the four organizations interacting together permitted the Navy to initiate and implement aviation innovation. This discussion might be abstract if it did not include examples of self-correcting organizations. Yet, the multiorganizational system of the interwar US Navy involved in aviation issues-the General Board, the Fleet, the Naval War College, and the Bureau of Aeronautics-actually engaged in self-correcting behavior. Recognition of the self-correcting and self-evaluating features of the relationship among those organizations is possible only when the interaction among the different levels of analysis is understood.

Policy makers who wish to foster innovation should devote attention to coordinating the interactions among agencies that will help senior decision makers learn about and anticipate problems and options. A self-correcting multiorganizational system can be created consciously, directing managerial attention to each of the three levels of analysis. By setting institutional rules for the continuous and rapid interaction of diverse and independent groups and agencies having different agendas, policy makers will receive a fairer hearing of the information and analysis needed to make decisions. Procedures can be designed and set in place to avoid common organizational maladies (such as goal displacement and uncertainty absorption) by focusing on organizational structure, veto points, and administrative redundancies. The ability of individuals to handle political and technological uncertainties can be enhanced through professional military education. A second order effect of tailoring adjustments to different levels of analysis would be the formulation of more cogent and convincing arguments to protect innovation from potential bureaucratic opponents.

This study does not employ a simple form of causal analysis, assuming a linear relationship between organizational structure and the organizational outcomes of adopting and implementing innovations. Innovation may be a nonlinear behavior-when viewed from the perspective of interactions among institutions, organizations, and individuals. If so, it is necessary to account for the influence of institutions and organizational structure on the innovation process because, as we have learned from studies of chaotic behavior and nonlinear dynamics, even "small errors of observation in the starting position may lead to virtually total unpredictability after some period of time."[43] But, the possibility that the innovation process may be nonlinear neither militates against the design of better management practices nor more profound generalizations concerning how decisions are made. The failure of current intellectual conceptions to explain revolutionary innovations reflects the poverty of thought-not an in-principle obstacle to gaining knowledge. For example, over the last 10 years ecologists have begun using nonlinear dynamics and chaos theory to interpret key features of population fluctuations, and their efforts have met with success. Recent laboratory studies have shown how chaotic and nonlinear behavior may be both modeled and predicted accurately.[44]

But, the weapons acquisition process is not random. Many individually simple rules regulate the actions of people. The passage of time alters the effect of individual and simple rules. The innovation process develops complex and unpredictable patterns due to the interaction of multiple actors and organizations, the success (or failure) in the evolution of technology, and the behavior of actors and organizations outside the system (e.g., potential adversaries or rivals). Despite simple rules, the number of actors and organizations and associated interactions make the system very complex. This complexity sensitizes the weapons acquisition process to chance actions.

The potential for the analytical strategy presented in this chapter becomes even clearer in the following two chapters, which describe the innovation process involved in the Air Force's B-52 development program. The case material will begin to illustrate the interaction among levels of analysis and the factors discussed in chapter 2-that is, chance, short time horizons, organizational structure, and poor decision making.

Notes

1. Arthur L. Stinchcombe. Constructing Social Theories (New York: Harcourt, Brace & World, Inc., 1968), 47-53.

2. Discussion of levels of analysis is a rich subject in philosophy of science and methodology. Physics Nobel laureate Philip W. Anderson's statement on the topic is admirably short and quite clear. See Anderson, "More is Different," Science 177, no. 4047 (4 August 1972): 393.

3. Douglass C. North, Institutions, Institutional Change and Economic Performance (New York: Cambridge University Press, 1993), vii, 3-5.

4. Ibid., 93-94. 5. I am indebted to Walton S. Moody for bringing this example to my attention.

6. David Sheff, Game Over: How Nintendo Zapped an American Industry, Captured Your Dollars, and Enslaved Your Children (New York: Random House, 1993).

7. Stephen Manes and Paul Andrews, Gates: How Microsoft's Mogul Reinvented an Industry and Made Himself the Richest Man in America (New York: Touchstone, 1993).

8. North, 9.

9. See, for example, Charles E. Lindblom, Politics and Markets: The World's Political Economic Systems (New York: Basic Books, 1977), 161-88. The total number of players in defining and setting military and acquisition policy during the interwar period was smaller in Britain than in the United States. The number of senior civilian players was small too. For most of the years between 1919 and 1939, the Committee of Imperial Defense had only five members. Most of the committee members were drawn from the aristocracy. See Sir Ivor Jennings, Cabinet Government, 2d ed. (Cambridge: Cambridge University Press, 1951), chap. 10. The patterns set by British formal and informal institutions limit access of a wide selection of people to government decision-making and affect the makeup of contemporary governance. Even standardizing for size of government, there is less access to the decision making process in Britain than in the United States. Heclo and Wildavsky, for example, note that governmental decisions are made by "small groups of powerful men who depend on each others' good opinion." Hugh Heclo and Aaron Wildavsky, The Private Government of Public Money: Community and Policy Inside British Politics (Berkeley: University of California Press, 1974), xvi.

10. North, 83, 89.

11. Robert Schlaifer, Development of Aircraft Engines (Boston: Graduate School of Business Administration, 1950), 59-60.

12. Thomas C. Hone and Mark D. Mandeles, "Interwar Innovation in Three Navies: USN, RN, IJN," Naval War College Review 40, no. 2 (Spring 1987): 63-83; Norman Friedman, Thomas C. Hone, Mark D. Mandeles, The Introduction of Carrier Aviation into the U.S. Navy and Royal Navy: Military-Technical Revolutions, Organizations, and the Problem of Decision (Washington, D.C.: OSD/NA, 1994), 4-5.

13. Tacit knowledge is acquired by practice and can be communicated only partially. The ability to acquire tacit knowledge is not distributed evenly throughout society or particular organizations. In contrast, communicable knowledge can be transmitted from one person to another. See Michael Polanyi, The Tacit Dimension (Garden City: Doubleday-Anchor. 1967); and North, 94.

14. Friedman, Hone, Mandeles.

15. The great relative importance of leadership and organizational capabilities in the competition between technologies is illustrated by NASA during the 1960s. NASA administrator James Webb employed tacit political knowledge with great skill. Webb was "an inspired and inspiring leader. He could turn easily from subtle political negotiations with Congress to attending detailed technology reviews of a critical system at a NASA field center." Leonard R Sayles and Margaret K. Chandler, Managing Large Systems: Organizations for the Future (New Brunswick: Transaction Publishers, 1993), xviii.

16. Thomas K. Glennan Jr., Policies for Military Research and Development, P-3253 (Santa Monica: RAND Corporation. November 1965); Albert O. Hirschman and Charles E. Lindblom, "Economic Development. Research and Development, Policy Making: Some Converging Views," Behavioral Science, April 1962. 211-22; Burton H. Klein, What's Wrong with Military R and D? P-1267 (Santa Monica: RAND Corporation, 1958); Klein, The Decision-Making Problem in Development, P-1916 (Santa Monica: RAND Corporation, 1960); Klein. Policy Issues involved in the Conduct of Military Development Programs, P-2648 (Santa Monica: RAND Corporation, October 1962); and William Meckling, "Applications of Operations Research to Development Decisions," Operations Research. May-June 1958, 352-63; and E. G. Mesthene, The Nature and Function of Military R&D, P-2147 (Santa Monica: RAND Corporation, November 1960); Thomas K. Glennan, Jr., and G. H. Shubert, The Role of Prototypes in Development, RM-3467/1-PR (Santa Monica: RAND Corporation, April 1971); Thomas A. Marschak, "Strategy and Organization in a System Development Project." in The Rate and Direction of Inventive Activity: Economic and Social Factors; a Conference of the Universities-National Bureau Committee for Economic Research and the Committee on Economic Growth of the Social Science Research Council (Princeton: Princeton University Press, 1962); Andrew W. Marshall and William H. Meckling, Predictability of the Costs, Time, and Success of Development. P-1812 (Santa Monica: RAND Corporation, October 1959), reprinted in The Rate and Direction of Inventive Activity: Economic and Social Factors (Princeton: Princeton University Press, 1962): Richard R. Nelson, "Uncertainty, Learning, and the Economics of Parallel Research and Development Efforts," Review of Economics and Statistics, 1961, 351-64; and Richard Langlois, "Industrial Innovation Policy: Lessons From American History," Science, 18 February 1983. 814-18; Robert L. Perry, The Mythography of Military R&D. P-3356 (Santa Monica: RAND Corporation, May 1966): Innovation and Military Requirements: A Comparative Study, RM-5182-PR (Santa Monica: RAND Corporation, August 1967); Giles K. Smith, Alvin J. Harman and Susan Henrichsen, System Acquisition Strategies, R-733-PR/ARPA (Santa Monica: RAND Corporation, June 1971); and American Styles of Military R&D, P-6326 (Santa Monica: RAND Corporation, June 1979).

17. North, 99.

18. I am indebted to George E. Pickett Jr. for bringing this example to my attention. Guy Hicks and George Pickett, "Airland Battle, Helicopters and Tanks: Factors Influencing the Rate of Innovation," unpublished paper, 9 August 1988.

19. Dwight D. Eisenhower, At Ease: Stories I Tell to Friends (Garden City, N.Y.: Doubleday & Co., Inc., 1967), 173.

20. Roger H. Nye, The Patton Mind: The Professional Development of an Extraordinary Leader (Garden City, N.Y.: Avery Publishing Group, 1993),43.

21. Martin Blumenson, ed., The Patton Papers 1885-1940 (Boston: Houghton Mifflin Co., 1972), 659-60,716-28; Nye, 43-52.

22. Nye, 52.

23. Blumenson, 716-20. See also Allan R Millett and Peter Maslowski, For the Common Defense: A Military History of the United States of America (New York: Free Press, 1984), 380-82.

24. This point is similar to the criticism of histories written by advocates of policy alternatives not chosen. See Nelson W. Polsby, Political Innovation in America: The Politics of Policy Initiation (New Haven: Yale University Press, 1984),77.

25. Millett and Maslowski, 378-82.

26. Friedman, Hone, and Mandeles, 65.

27. Historians studying military innovation have not recognized the role of such evaluations and interactions in multiorganizational systems. As a result, their explanations of events are flawed and their policy-related proposals are of limited utility to policy makers. For example see the unidimensional analyses in Allan R Millett and Williamson Murray, eds., Innovation in the Interwar Period (Washington, D.C.: OSD/Net Assessment, June 1994), especially Geoffrey Till's "Adopting the Aircraft Carrier," and Williamson Murray's "Innovation: Past and Future." See also Millett and "," Maslowski, 382.

28. The interaction about the concept and technology of carrier aviation among separate organizations in the interwar Navy seems analogous to the type of interaction Undersecretary of Defense Paul G. Kaminski is trying to institute with the concept of "integrated product teams." See chapter 6 for a theoretical justification of Kaminski's acquisition procedure concept.

29. Michael D. Cohen, James G. March, and Johan P. Olsen, "A Garbage Can Model of Organizational Choice," Administrative Science Quarterly, 1972, 1-25. See also James G. March and Johan P. Olsen, "Garbage Can Models of Decision Making in Organizations," in Ambiguity and Command: Organizational Perspectives on Military Decision-Making, ed. James G. March (Boston: Pitman, 1986); Karl E. Weick, "Educational Organizations as Loosely Coupled Systems," Administrative Science Quarterly, 1976, 1-19; Karl E. Weick, "Sources of Order in Unorganized Systems: Themes in Recent Organizational Theory," in Organizational Theory and Inquiry, ed. Yolanda S. Lincoln (Beverly Hills, Calif.: Sage Publications, 1985): John W. Kingdon, Agendas, Alternatives, and Public Policies (Boston: Little, Brown & Co., 1984).

30. James G. March, "Ambiguity and Accounting: The Elusive Link Between Information and Decision-Making," in Decisions and Organizations, ed. James G. March (New York: Basil Blackwell Inc., 1989), 390-91.

31. John P. Crecine, "Defense Resource Allocation: Garbage Can Analysis of C3 Procurement," in Ambiguity and Command: Organizational Perspectives on Military Decision-Making, ed. James G. March {Boston: Pitman, 1986), 84-6.

32. Roger Weissinger-Baylon, "Garbage Can Decision Processes in Naval Warfare," in Ambiguity and Command: Organizational Perspectives on Military Decision-Making, ed. James G. March (Boston: Pitman, 1986), 38.

33. March, "Introduction: A Chronicle of Speculations About Decision-Making in Organizations," 8.

34. Ibid., 7, 13.

35. Crecine, 84-5.

36. March, "Introduction: A Chronicle of Speculations About Decision- Making in Organizations," 13: March, "Ambiguity and Accounting," 390-91.

37. Cohen, March, and Olsen, "A Garbage Can Model of Organizational Choice," and March and Olsen, "Garbage Can Models of Decision Making in Organizations." Also see the other chapters in Ambiguity and Command.

38. For example, Polsby notes that "Whoever put the words 'maximum feasible participation' into the community action legislation seems to have done his work so unobtrusively that a few years after the event, nobody is quite sure who did it or how it was done." Polsby, 151.

39. Stephen P. Rosen, Winning the Next War: Innovation and the Modem Military (Ithaca, N.Y.: Cornell University Press, 1991), 187,250,254.

40. E. E. Schattschneider, The Semisovereign People: A Realist's View of Democracy in America (Hinsdale, Ill.: Dryden Press, 1975).

41. Friedman, Hone, and Mandeles, chap. 5. 42. Martin Landau, "On the Concept of a Self-Correcting Organization," Public Administration Review 33, no. 6 (November/December 1973): 533-42. Aaron Wildavsky, "The Self-Evaluating Organization," Public Administration Review 32, no. 5 (September/October 1972): 509-20.

43. Kenneth J. Arrow, "Workshop on the Economy as an Evolving Complex System: Summary," in The Economy as an Evolving System, ed. Philip W. Anderson, Kenneth J. Arrow, and David Pines (Redwood City: Addison-Wesley Publishing Co., Inc., 1988), 278, 280.

44. Peter Kareiva, "Predicting and Producing Chaos," Nature, 18 May 1995, 189-90. See also R. F. Costantino et al. "Experimentally Induced transitions in the Dynamic Behavior of Insect Populations," Nature, 18 May 1995, 227-30.

Chapter 4
Prelude: Jet Propulsion and the Air Force

[T]he Air Corps does not, at this time [1941], feel justified in obligating funds for basic jet propulsion research and experimentation.

-Brig Gen George H. Brett

The Army's organization and its senior officers' approach to learning about tanks and associated operational concepts, described in the last chapter, also characterized its approach to advancing aviation technology and. in particular, jet propulsion. The concept of jet propulsion was known for hundreds of years before it was put into practice. Gas turbines were run in France in 1906 and the United States in 1907. In 1921 the first patent for a complete turbojet was filed in France. [1] Yet, the Army's acceptance of the turbojet concept as a means of high-speed propulsion had to wait for particular social and organizational conditions, inventions in aerodynamics, metallurgy, and chemistry, as well as war.

These next two chapters examine the difficulties encountered by the interwar Army in relation to the adoption of jet propulsion. The origin of these difficulties lay in a post-World War I Army that was neither created nor designed to encourage and adapt to technological change. [2] Rather, the Army leadership attempted to employ existing capabilities on known and understood mobilization, tactical, and logistical problems. [3] Although the presence of a growing knowledge base and variety of technologies in the interwar period created the potential to innovate, the Army leadership saw no need to create and maintain an organizational potential to invent new things or identify and accomplish new tasks. Army leadership, accustomed to issuing orders, was ill-prepared for the many situations, conditions, and events associated with turbojet technology that were outside their vision or control. As James G. March. Stanford University's Jack Steele Parker Professor of International Management, has noted,

> decision makers who have little experience with observing. understanding. and reacting to changes in the environment lose the capability to do so. Their power to impose a world undermines knowing how to cope with a world that cannot be unilaterally controlled. As their skills at imposing environments grow, their skills at adapting to an environment atrophy. [4]

Compounding this common pitfall of leadership and command was the impact of the organizational and institutional setting on decision-making. As argued in chapter 3, few Navy or civilian officials concerned with developing new military capabilities in the interwar period understood how interactions among relevant organizations-the General Board, the Fleet, the Naval War College, the Bureau of Aeronautics, and Congress-could enhance the intelligence of their decisions about aviation-related technologies.

Such multiorganizational systems are a key to more effective-that is, smarter-policy making. Multiorganizational systems typically feature overlap of policy concerns among pertinent organizations and an absence of strong formal coordination mechanisms for policy making. The Navy's General Board, for example, did not have formal authority to mandate compliance with its decisions. However, the high prestige of board members-senior admirals-was enhanced by an analytic and fact-finding style widely seen as fair. These conditions helped ensure coordination of independent agencies to create the required information through wargames, fleet maneuvers, and simulations.

Through this ad hoc multiorganizational system, the Navy displayed robust informal and formal lateral communication and coordination as independent organizations dealt with a common core problem or set of tasks. The results of negotiation and discussion among the various Navy organizations involved in carrier aviation development was a self-organizing and self-regulating system that exhibited coordinated action "when and where required, rapid response, sustained improvement, [and] high reliability." [5]

Effective functioning of a multiorganizational system may be sensitive to types and amounts of overlap or outright duplication. The Navy aviation community's effective decision making about technological innovation began to decline as post-World War II organizational and administrative changes were made to produce a more "streamlined" administrative system having less overlap and duplication. [6] As we see in this and the next chapter, the extent to which the Army, and later the Air Force, permitted overlap, duplication, and a multiorganizational approach directly affected their ability to develop effectively the potential of turbojet propulsion. Five interdependent issues shape the particular environment that influenced turbojet technology from the 1930s to late 1940s.

• The operational requirement. What were the origins of the turbojet operational requirement? How did senior leadership evaluate the value of turbojet propulsion as a means to ensure the accomplishment of strategic roles?

• The operational concept. What was the initial concept of operation? How did senior leadership view the role of airpower in combat and what did they believe turbojet propulsion could do to fulfill that role?

• The technology. How did the turbojet technology evolve? What technical developments made turbojet propulsion possible?

• The organization. What organizational and bureaucratic factors influenced the adoption of turbojet propulsion? What role did organizational structure play in the adoption of turbojets?

• Implementation. How was the decision to accept turbojet propulsion implemented? What tasks had to be accomplished to get the systems accepted and into service?

Discussion of these questions provides a context for the next chapter's detailed examination of jet propulsion and bombardment aircraft acquisition decisions.

Origins of the Jet Propulsion Operational Requirement: Interwar through Post-World War II

The first US military operational requirement for jet propulsion in 1941 was not the result of a conscious search for alternative means of propulsion for high-speed flight. [7] Rather, it came from Gen Henry H. "Hap" Arnold's chance introduction to the Whittle engine. Arnold's personal role in establishing the military value of turbojet propulsion was necessary because the Army Air Corps was not organized to propose, to learn about, or to evaluate new and alternative technologies and associated operational concepts. Indeed, Arnold himself had been blind to the potential of high -speed jet propulsion until March 1941, when he saw an operating engine in England. Recognition of jet propulsion's potential was not a case of one man's prescience or intuition in the face of official indifference. Instead, it was a story of whether an organization-the Army Air Corps-was structured to enhance the intelligence of its members about technological and acquisition matters.

Through the 1930s, there was no choice between propeller and jet propulsion; civilian and military aircraft were powered only by reciprocating propeller-driving engines. Some engineers were aware that turbojet propulsion could be a useful alternative to reciprocating engines. In 1919, years before they were practicable, turbojets had been proposed as power plants for aircraft in the United States and in Europe. [8] In the early 1920s, French and British experimental work on jet propulsion was published. In 1922 the US Bureau of Standards investigated the turbojet as a means of aircraft propulsion. Two years later, the Bureau of Standards' investigator, Edgar Buckingham, concluded that jet propulsion would be impractical for either civilian or military purposes: the top speed of a jet-powered aircraft would be only 250 MPH, fuel consumption would be four times higher than piston engines, and the turbojet engine would be more complicated than a piston engine. [9]

In contrast to these perceived limits of turbojet propulsion, propeller-driven aircraft performed acceptably, were affordable in quantity, and had extensive manufacturing facilities, as well as known maintenance and repair procedures. The sheer number of cheap World War I "Liberty" reciprocating engines available also militated against the appearance of a market-initiated turbojet. [10] These factors led to a "lock-in" (as described in chapter 3) of reciprocating engines in civilian and military aircraft, and created insurmountable obstacles for the inventors and would-be popularizes of alternative power plants. Power plant engineers considered aircraft design and the speed of propeller-driven aircraft as givens; aircraft engines and their performance were investigated in terms of existing speed ranges. [11]

While proposals for jet aircraft reappeared persistently in the United States, subsequent studies completed at the Bureau of Standards and National Advisory Committee for Aeronautics (NACA) confirmed Buckingham's 1924 conclusions that low speeds and high fuel consumption made jet propulsion impractical. By the mid-1930s, however, some researchers in the aircraft design community were beginning to question the possibility of continued progress in propeller-driven aircraft technology. In 1934, John Stack, a NACA researcher, reported a practical limit on speeds to be achieved by propeller-driven aircraft. A different power plant would have to be employed if aircraft were to fly near or faster than the speed of sound. [12]

In Europe, scientists and engineers who were investigating alternatives to propeller-driven aircraft gathered in 1935 at the Volta conference on high -speed aircraft. Sponsored by the Italian Academy of Sciences and held at Rome, the Volta conference brought together the world's preeminent aerodynamicists, including the Hungarian-American Theodore von Kármán. He was impressed with the reported research results on high-speed flight and concerned with the poor position of US theoretical research. While American engineers produced first-rate empirical design data for subsonic aircraft and for reciprocating propeller-driven aircraft, German or German-educated scientists led the theoretical investigations of high -speed and turbocompressor phenomena. In terms of quality of theoretical research on high-speed flight, the British lagged slightly behind the Germans, and American, Italian, and French scientists lagged badly. [13]

Not only were Europeans doing the best theoretical research, they were building the research tools to maintain that research lead. While attending the conference, von Kármán visited the Italian research center at Guidonia, and saw an Italian 2,500-MPH wind tunnel that was used to investigate supersonic phenomena. Upon return to America, von Kármán attempted to convey the need to engage in theoretical high-speed research, and proposed to the Air Corps leaders that they build a supersonic wind tunnel. Air Corps officers rejected the idea, arguing that a supersonic tunnel would be too expensive and, in any event, discretionary money was unavailable. Von Kármán also urged NACA to build a supersonic wind tunnel. But George W. Lewis, NACA's executive director, could not understand the need to build a tunnel capable of speeds greater than the existing NACA 650- MPH tunnel because propellers rapidly lose efficiency at speeds greater than 600 MPH. Lewis did not examine the assumption that aircraft would always rely on propellers, and no multiorganizational system existed to raise doubts about or to challenge his decision. [14]

The first workable American turbojet capable of propelling a military aircraft was based on a model designed by Frank Whittle of Great Britain, the model that so impressed Hap Arnold in 1941. Whittle faced great obstacles to receiving support for an experimental program. He had patented a design for a turbojet engine in 1930 and attempted to interest the British Air Ministry in the idea. But officials regarded turbojets as impractical and would not provide funding. Due to lack of official interest, Whittle allowed his patent to lapse. A chance encounter with two retired Royal Air Force (RAF) officers and subsequent contacts led through a chain of personal acquaintances to investment bankers, O. T. Falk & Partners, Limited, who put up the first capital to back Whittle's work. One member of the investment firm had scientific training and was open to the possibility of jet propulsion. However, he underestimated how much money would be spent before governmental support would be obtained and a flyable engine built. [15] In 1935 Whittle opened Power Jets, Ltd., to develop a turbojet engine. By 1937 a bench model of the turbojet was operating. It demonstrated the feasibility of the turbojet concept in terms of thrust-to-weight ratio and fuel consumption. [16]

With these promising results, Whittle again approached the Air Ministry and was referred to the engineers at the Royal Aircraft Establishment at Farnborough (the British equivalent of Wright Field). Air Ministry officials rejected Whittle's proposal a second time. Chance, however, again intervened in the treatment of Whittle's idea through the presence of Sir Henry Tizard as chairman of the Air Ministry's engine subcommittee of the Aeronautical Research Council. Tizard liked Whittle's proposal and persuaded the ministry to give Whittle financial support. [17]

In the late 1930s, while Power Jets and the Air Ministry were taking these first hesitant steps toward designing and building a working turbojet engine, leaders of the US Army Air Corps still did not anticipate the need for jet propulsion, did not initiate an aggressive development program, and had no multiorganizational system like the Navy's to guide aircraft acquisition. The intellectual interaction among members of the Navy's aviation community and other parts of the Navy was not duplicated by the Army. Senior Air Corps leaders saw jet research as a long-range project, and leaders of engine firms saw no reason to conduct earnest investigations of turbojet propulsion. [18] The earliest complete designs and serious proposals for development of gas turbines came in 1941 from such airframe builders as Northrop and Lockheed. [19]

In 1940 the Air Corps tried to get the National Defense Research Committee (NDRC) to assume responsibility for the entire jet propulsion program. NDRC officials, however, argued that the responsibility for jet research resided with NACA. [20] NACA officials responded that it was not the agency primarily responsible for engine development. Eastman N. Jacobs, a NACA researcher, had done some work on jet propulsion in the early 1930s-and again in 1938-39-but NACA's leadership did not appreciate the importance of the topic. [21] In February 1941, the Air Corps asked NACA to establish a Special Committee on Jet Propulsion. [22] Chairman William F. Durand convened committee in March 1941. [23] General Arnold opposed including reciprocating engine manufacturers on the committee for fear they would divert money and personnel away from producing needed piston engines. The three companies represented on the committee, Allis-Chalmers, Westinghouse, and General Electric Company (GE), were manufacturers of turbines. They chose to study axial rather than centrifugal compressors, because of NACA results that showed that higher efficiencies could be obtained with axial compressors. In September 1941, the Durand committee recommended that development contracts be given to each of the three companies. [24]

Meanwhile, D. R. Shoults, an American technical representative for GE turbo superchargers (used in the B-17's engines) in Britain discovered that Frank Whittle was building a turbojet, and informed the US Army Air Corps technical liaison officer. Col A. J. Lyon. They obtained permission from the Ministry of Aircraft Production to conduct an inspection at Power Jets. When Arnold visited Britain in March 1941. Shoults and Lyon informed him of the technical progress. It was only then that Arnold realized that turbojets were being built, could be used for aircraft propulsion, and that the British would soon flight test a turbojet-powered aircraft. [25] Realizing the implications of turbojets, Arnold arranged to have a Whittle jet engine (designated the W-IX) and production drawings (for the W-2B) shipped to the United States for quantity production. [26] The urgency of his efforts increased after the first flight of a turbojet powered aircraft, the E28/39, on 15 May 1941. [27]

General Electric, because of its experience in building turbosuperchargers, was selected to build the American version of the Whittle engine. This development effort was shrouded in secrecy; the British opposed sharing design information with US firms beginning to investigate turbojet propulsion. [28] These secrecy requirements limited the flow of information and-because NACA was not permitted to participate in this research-the use of the best test equipment. [29] The engineers of important aircraft firms were surprised by progress in turbojet power plants when they were allowed to see the data. For example, in September 1944. Boeing engineers were surprised "that jet engines have developed to this stage; that a flying airplane such as the XP-59 has actually been built." [30]

It is interesting to contrast the effects in the United States of these information transfer restrictions with the policy within Britain where the Air Ministry promoted the fullest exchange of information among private firms and public agencies working in the field. [31] It was soon discovered that American firms operating independently and without outside help could not quickly catch up to British firms that had begun turbojet research years earlier and were continually sharing results. [32]

Whittle Engine

With hindsight, historians have criticized Air Corps leaders for their failure to pursue turbojet propulsion in the 1930s. Robert Schlaifer concluded that the lag in perfecting jet fighters was the "most serious inferiority" in American aeronautical development during World War II, an evaluation echoed by historians Alfred Goldberg, Irving B. Holley Jr., and Alex Roland. [33] In particular, Holley rejected the excuses offered by Air Corps apologists for the failure to initiate jet propulsion research. These excuses included scarcity of funds and faulty technological intelligence. Holley noted that Whittle's firm, Power Jets, Ltd., was undercapitalized and ill-equipped. Power Jets used a makeshift machine shop and, until 1939, received the equivalent of only about $5,000 from the Air Ministry. Between 1936 and 1939, Power Jets spent only about $100,000 to get a bench engine capable of showing the potential of turbojets. [34] With respect to technological intelligence, Holley asserted that there was ample public evidence of a real competitor to piston-driven propulsion at the 1935 Volta conference, and American engine manufacturers were aware of the potential competition from turbojets. Some American firms tried to translate this potential into reality. Wright Aero and Pratt and Whitney, and the airframe firms Northrop and Lockheed, initiated studies of turbine-powered aircraft. Air Corps leaders, however, were uninterested in these studies. [35]

For an explanation, we must look at the interwar years when Air Corps leaders were deeply involved in issues of doctrine, organization (including promotion policy and command and control of air units), and mission. With respect to doctrine, the first half of the 1930s had produced a conviction, on the part of Air Corps leaders, that independent strategic bombing operations could achieve "decisive" combat results and that airpower could prevent an armed invasion. US Air Corps Tactical School bombardment theorists argued that destroying the enemy's economic infrastructure would end enemy capability to wage war and, eventually, reduce enemy morale. The strategic bombing offensive would be implemented by precision bombing attacks on carefully selected economic-industrial targets. [36]

Not coincidentally, strategic bombardment- doctrine was a key element in the efforts of Air Corps leaders in the 1930s to create an independent Air Force, and organizational questions also occupied Air Corps leaders during the interwar period. [37] Throughout the 1920s, air units had been divided among ground unit Army commanders. Senior Air Corps leaders argued that a more efficient organization would centralize air units under a senior air commander in one General Head- quarters Air Force. The War Department permitted this reorganization, but would not agree to the formation of an independent Air Force or to its expansion at the expense of the Army.

Air Corps views about doctrine and organization tied neatly into the approved mission in national security: air defense of the United States and its overseas possessions. [38] Senior Air Corps leaders' preoccupation with advancing this doctrine and mission in support of an independent organization overshadowed questions of radical technological advancement. Time and attention-to the myriad issues facing Air Corps leaders-are neither free nor available in unlimited amounts. The difficult political problems regarding doctrine and missions required planning, arguing, and negotiation over a period of years. It is not surprising, as Holley asserted, that Air Corps leaders were ignorant of their technological options and made no effort to illuminate "unknown unknowns"-those matters about which they were both unaware and ignorant. Their emphasis on matters of coordination and competition with Army ground officers precluded focused attention on the problem of learning about and being able to evaluate state-of-the-art technology and aerodynamic theory. [39]

The tendency of senior Air Corps leaders to attend primarily to political issues in the interwar period was exacerbated by the Army's organization, which was not structured to direct attention to long-range research projects. Acquisition rules established in the 1926 Air Corps Act militated against experimental research. The Air Corps organization, like the Army General Staff, was suited to repetitive problems of training, equipping, and maintaining forces. It was ill-suited to arrange competing experimental analyses of unproven ideas about propulsion, aerodynamics, or operational concepts such as long-range fighter escort. Air Corps leaders neither sought, nor had available, a variety of ideas about new weapons technology or operational concepts. No multiorganizational system capable of examining future weapons or operational concepts existed, although such a multiorganizational system-composed of elements of the Air Corps, NACA, Army General Staff, War Department offices concerned with aviation, and analysts at the Air Corps Tactical School-could have been created.

These constraints continued to operate after World War II. Although the Army Air Forces (AAF) was unable to implement fully the strategic bombing offensive in World War II, it emerged from the conflict with high prestige, and its leaders were convinced of the effectiveness of the strategic air mission. But this prestige did not insulate Air Force leaders from difficult choices. Operational requirements for post-World War II strategic bombers were set in a political environment complicated by on-going arguments, negotiations, and trade-offs over doctrine, uncertainty about the employment of atomic weapons, competition from the Navy, and the struggle to create and build a new organization separate from the Army. Strategic bombardment doctrine and the development of long-range heavy bombers provided continuity and direction for air leaders, and it was assumed that propeller-driven aircraft would continue to be the means of strategic bombardment.

At the start of World War II, Air Corps leaders had recognized the range and payload limitations of their current aircraft. Therefore, in 1941 they contracted for development of an aircraft to provide a long-range bombing capability in the event overseas bases were denied to the United States through the defeat of Great Britain. This aircraft became the B-36 Peacemaker. [40] It marked a significant advance toward realizing the requirements that called for a 10,000-mile range with a 10,000-pound bomb load. [41]

The maiden flight of the XB-36 occurred in August 1946; the first production model (B-36A) was delivered in May 1948. On shorter missions in the 1,200 to 1,800-mile range, the B-36A could carry between 86,000 and 60,000 pounds of bombs (respectively), and thus would go far in fulfilling strategic bombardment doctrine. [42]

The requirements for a long-range bomber meshed nicely with the World War II analysis of Air Corps officers of the possibility of war with the Soviet Union and the need for basing rights. [43] After World War II, the War and State Departments' concern for overseas bases matched Air Force leaders' efforts to eliminate dependence on the Army and Navy to secure and maintain an overseas basing system in wartime. As the Air Force acquired targeting information on the Soviet Union, it opened an industry competition to develop anew, second-generation of heavy intercontinental bombers to reach these targets from the United States.

Military characteristics for a long-range bomber released in November 1945 called for an operating radius of 5,000 miles, a speed of 300 MPH at an altitude of 43,000 feet, a 10,000-pound bomb load, and maximum armor protection. There was little doubt among Army Air Forces' leaders that this aircraft would be a large straight-wing turboprop. [44] Boeing won the competition with Model 462: a large, tapered straight-wing' airplane powered by turbo-propeller engines with a radius of only 3,110 miles. Model 462 would later be designated the B-52.

B-36A Peacemaker

Articulating characteristics' for a long-range bomber was a messy and fluid process. Air Force planners saw little need to require turbojet propulsion, as the high fuel consumption of World War II-era jet engines made their use in long-range aircraft impractical. Wartime requirements for a medium turbojet-powered bomber, released in June 1943, exceeded the state-of-the-art and included a range of 3,500 miles, a service ceiling of 45,000 feet, and an average speed of 450 MPH. [45] There was no real hope that a jet engine could be matched to a large airframe capable of carrying a heavy payload on an intercontinental flight. [46]

Internal political trade-offs heavily influenced the formulation of new aircraft operational characteristics, and exploitation of new technology was not a major concern. Air Force leaders had three more pressing concerns: to avoid dependence on the Army, Navy, or foreign allies; to justify strategic bombardment doctrine; and to enhance the Air Force's role in national security mission and budget debates. The Air Force's development of post-World War II bombardment aircraft, such as the B-52, was greatly affected by the political bargains inherent in these three goals. The many restatements of military characteristics throughout the development of the B-52-in a search for adequate aircraft performance-reflected uncertainty over the ability of the B-52's paper design to meet the Air Force's political needs. There were considerable design changes over a three-year period. The initial operational requirements, won by Boeing, did not specify a jet-powered aircraft. The final aircraft-swept-wing aircraft propelled by eight turbojets-was very different from the tapered straight-wing turboprop accepted initially, and it was not the result of Air Force leaders' prescient understanding of technological evolution.

Initial Concept of Operation

The B-52's initial operational concept as a platform for delivery of atomic bombs "as soon as hostilities start" was based on World War II bombing campaigns and the need to base the aircraft in the United States. Air Force leaders already knew how to design tactics and to conduct air warfare; they planned to attack the economic and military targets that sustained the enemy's military forces. [47] Thus, they sought to fit new aeronautical technology and the characteristics of new aircraft and munitions into their existing conception. The use of the B-29 during World War II, for example, provided the model for employment of a medium bomber. These aircraft would be utilized after the B-52s delivered atomic strikes. A coherent analysis of how long-range bombers would be used in the future was neglected as other issues diverted the attention of Air Force leaders, including the numbers and physical characteristics of atomic weapons, aircraft characteristics trade-offs, foreign policy interests, and interservice rivalry with the Navy. In the end, preoccupation over these matters only reinforced the main concerns of Air Force leaders that air operations should be run by an independent service and that the Air Force should be accorded the premier role (and budget share) in national security. As a consequence, they failed to search coherently for operational concepts or to estimate the consequences of acquisition policies on future war-making

capabilities.

The scarcity of atomic bombs placed a high political premium on convincing important decision makers outside the Air Force that long-range land-based bombers were the optimum delivery vehicle, and that sharing scarce weapons with the Navy would detract directly from the Air Force role in atomic warfare. By June 1948, the atomic bomb production rate had reached two per month, up from one every two months in fiscal year 1947, but the number of completed weapons in the nation's stockpile was probably less than 50. [48]

Within the Air Force, secrecy concerning the stockpile and technical characteristics of the weapons complicated the design of a nuclear-capable force. The Air Force wanted light weapons, but the Atomic

Energy Commission (AEC) did not release specific information about weight. [49] Hence, the B-52 bomb bay design remained open throughout 1948 to provide for the possibility of a 15,000-pound bomb instead of a 10,000-pound bomb. [50] The design bomb weight was not reduced to 10,000 pounds officially until mid-January 1949. [51] In the meantime, the additional 5,000 pounds reduced the B-52's projected range and raised the possibility of costly changes in bomb bay configuration.

The range-speed trade-off was a critical issue throughout the B-52's development. Claims used to support any particular range-speed trade-off could not be evaluated rigorously through tests like those conducted on particular aircraft components by engineering officers at Air Materiel Command (AMC). The proper range-speed trade-off for mission success could be settled only in an actual combat campaign against current (and extrapolated future) Soviet ground- and air-based air defenses. The absence of an unambiguously correct range-speed compromise for the B-52 allowed conflicting strong beliefs about military operations for the airplane. Maj Gen Curtis E. LeMay's views of the requirements for an intercontinental bomber, for example, differed substantially from those advanced by AMC officers. Despite the small number of atomic bombs in the stockpile, LeMay wanted an aircraft whose primary mission would be to engage in long-range atomic warfare. He recognized that the ability of the Air Force to perform an atomic attack, although unstated in the 1946 conferences on B-52 characteristics, would help the Air Force argue for its own role in the post-World War II organization of national security.

In contrast, AMC officers advocated aircraft requirements that were more easily achievable, such as an advanced B-50 capable of delivering conventional high-explosive bombs. Yet without the threat of a "hot" war, senior Air Force officers had little incentive to reach for this more easily achievable technical solutions to the military requirement for long range.

American foreign policy supported the atomic component 0 the B-52's operational concept. "Containment" was adopted in 1947 as the basis of foreign policy, and the Truman administration was moving to a military strategy of nuclear deterrence. The concept of deterrence received high level Ai Force sanction in January 1948 when the President's Air Policy Commission, led by Thomas K. Finletter, concluded that national security depended upon the prospect of a counterattack of utmost violence to any attacking count!) Strategic deterrence became the formal doctrine when the National Security Council (NSC) drafted NSC 20/4 in November 1948. This document stated United State objectives in relations with the Soviet Union: United State security rested on maintaining military readiness as long a necessary to deter Soviet aggression. [52] The American policy, deterrence reinforced the need perceived by Air Force Headquarters officers for a long-range bomber to fulfill the A Force strategic mission.

As the nation's postwar defense establishment struggled over roles and missions, the Air Force's conflict with the Navy climaxed, temporarily, in the B-36 congressional hearings. [53] The Navy had become interested in the offensive use of atomic weapons in 1947. This interest manifested itself in Navy proposals for funds to construct a new large flush-deck aircraft carrier, which would be capable of atomic warfare with appropriate aircraft. By 1946 Naval aviation officers had understood that the combination of jet engines and atomic weapons opened the "possibility of adding true strategic strike capability to [the Navy's] other offensive roles, since it was no longer necessary ...to think in terms of giant [land-based] superbombers for strategic operations." [54] The Bureau of Aeronautics outlined a proposal for a bomber capable of operating from the large flush-deck carriers being planned. Douglas Aircraft Company received the contract for what became the jet-propelled A3D Skywarrior, and a suitable design for the aircraft was completed in 1949. The Navy aircraft requirement was met by a swept-wing jet bomber, weighing 60,000 pounds (the largest and heaviest then projected for use), with a large internal weapons bay with provision for 12,000 pounds of conventional or atomic bombs. [55] A briefing prepared at AMC's Bombardment Branch noted Navy plans to develop a bomber that could compete with the B-52. Since the Navy airplane would be carrier based, its required radius could be shorter than continental US-based aircraft. Even so, the Navy airplane's maximum speed of about 600 MPH, ceiling of 41,000 feet, and radius of more than 1,000 miles exceeded the performance of the Air Force's March 1948 version of the B-52 (Boeing Model 464-35).

Throughout 1948 Air Force Secretary W. Stuart Symington and top Air Force officers worked to strengthen public and congressional support for Air Force positions. Headquarters officers were convinced of the need to acquire the very best aircraft to replace the B-36: anything less would abet the critics who wished to harm the Air Force's public image and retard achievement of Air Force goals, for example, a 70-group Air Force. As the decision process concerning the B-52's initial operational concept became diverted by interservice rivalries, uncertainties about atomic technology, and conflicts over mission, the Air Force failed to advance the conception of the B-52's bombardment operations much beyond the bombing doctrine associated with bombers already in the inventory.

Evolution of Technology

A properly designed propeller can satisfy almost any speed and load conditions of an airfoil efficiently, providing the velocity of the airfoil and of the propeller tips do not reach the speed of sound. As airfoil speeds approach Mach 1, propeller efficiency decreases quickly. Propeller efficiency becomes extremely low at an airfoil velocity of 500 MPH. The altitude at which propellers will function efficiently also is limited. In propeller-driven aircraft, the propeller captures and accelerates air, pushing it backward and thrusting the aircraft forward. At higher altitudes, the atmosphere is less dense and, hence, the propeller gets less "bite." The absolute ceiling for propeller-driven aircraft is approximately 55,000 feet. [56]

The turbine in a turbine-propeller (turboprop) configuration supplies power to the compressor and rotary power to turn the propeller. The main disadvantages of propeller systems are their weight and complexity. [57] Yet, turboprop engines are still more efficient than turbojets at speeds ranging from 300-400 MPH and altitudes of 30,000 to 40,000 feet. In contrast, turbojets become efficient at airfoil speeds of 400 MPH; 80 percent of a turbojet's power is wasted at 300 MPH. Turbojets also can operate at higher altitudes than propeller-driven aircraft; they begin to press their absolute ceiling at about 80,000 feet; the less dense air at higher altitudes produces less drag and the colder temperatures improve compressor efficiency.

Turbojet propulsion uses a gas turbine to power an air compressor; it produces propulsive thrust solely by the ejection of a high-velocity gas stream through a nozzle. A turbojet uses the atmosphere both as its operating medium and as source of the oxidizing agent for its fuel. [58] The basic principle of jet or reaction propulsion is to heat a gas within the engine at greater than atmospheric pressure and direct the heated and compressed gas to the atmosphere at the rear of the engine. The gas expands as its pressure falls, increasing the velocity with which it is expelled from the engine. The reaction to the forces expelling the gas from the engine is the thrust that drives the airfoil forward. [59] A gas turbine or jet engine has three major parts. The compressor section contains one or more sets of blade-tipped wheels that suck in and compress air. The burner section burns fuel using oxygen from the atmosphere. The turbine section drives the compressor, allowing the unusable energy from the hot gases to exit the engine and provide thrust. [60]

Beginning in the late 1930s and continuing through early 1941, GE and the Army traded bulletins and technical papers about gas turbines. In 1940 two GE engineers produced a paper entitled" Airplane Gas Turbine with a Propeller or Jet Propulsion," with completed curves for different altitudes and pressure ratios for turboprop and turbojet propulsion systems. In the meantime, GE engineers saw Eastman Jacobs's work at NACA on axial flow gas turbines. Up until this time, most research on gas turbines had been based on the centrifugal flow compressor principle. Centrifugal flow compressors whirl incoming air in a circular casing, compressing it into higher pressure by forcing it to the outside of the casing. Designers realized that there was a crucial trade-off between engine power and the centrifugal compressor's size and weight. Furthermore, not only would the centrifugal compressor be too heavy, but the frontal area of the engine would present too much drag.

The axial compressor, in contrast, uses a series of blades attached to a rotating shaft combined with blades fixed in the casing to compress air as it passed through an enclosed tube. GE engineers believed that the axial compressor answered critical design questions about the turboprop power plant and turbines capable of powering Navy PT boats. [61]

Initial wartime US military interest in turbojet power plants centered on the Whittle engine. Arnold chose GE to conduct jet engine research-and build 15 engines-because of its long experience on turbosuperchargers and research on turbine technology. Within six months of the contract, the company began testing its version of the Whittle W-2B production engine in March 1942 and had a working engine in April. [62]

On 2 October 1942, a Bell XP-59A Airacomet, powered by two GE (Type I-A) jet engines, flew for the first time. Flight testing continued for about a year. In July 1943, GE's material and design changes led to a new engine, the I-16. Two I-16s installed in an XP-59A propelled the aircraft to an altitude of 46,700 feet. By June 1944, GE had produced several improved versions of the Whittle engine, including the I-14, I-16, and I-20.

GE also was developing 4,000-pound thrust engines, later designated the J33 and J35. A J33 engine powered the Lockheed XP-BOA Shooting Star on its first flight on 10 June 1944. [63]

Lockheed XP-80A Shooting Star

Despite advances made during World War II, the field of aerodynamics remained undeveloped. [64] There were, however, high hopes for the potential of new types of power plants. The Army Air Forces' 1943 request for proposals on medium-range jet bombers was revealing some interesting results. For example, Boeing officials believed that the straight-wing airframes being examined did not match the performance potential of the jet engine. They asked permission to study the problem further. [65] In late October and early November 1944, Boeing officials presented the results of preliminary studies to the AAF. The chief of engineering division at Wright Field summarized conclusions from those discussions:

> Contrary to the existing belief that jet-propelled aircraft are limited to short ranges and fighter-type aircraft, the Boeing studies indicate that ...the range of jet-propelled aircraft may exceed that of conventional aircraft. Also, the studies indicate that if a speed of approximately 600 mph is attained, the specific fuel consumption of jet-propelled aircraft may approach that of conventional airplanes. This opens up the attractive possibility of developing at least medium-range, jet-propelled bombardment airplanes. ...As a result of these discussions and Boeing studies. it appears to be entirely feasible to build an airplane which will meet all of the desired performance requirements. [66]

Some propulsion experts believed that swept-wing platforms and jet and rocket engines would result in aircraft capable of much higher speeds. [67] Yet, there was little experience and knowledge about transonic and supersonic aerodynamics. [68] Nor was there good theoretical direction or experience in wing design for transonic and supersonic airplanes. [69] In May 1945, the Boeing medium bomber design team was directed to review captured German research data on the design of swept wings. These data solved some problems related to aircraft speed, but raised new questions. [70] For example, the Boeing B-47 medium bomber's swept-wings introduced new problems of stability, control, aeroelasticity, flutter, strength, and performance. No one even knew how to mount turbojets onto swept wings. [71] Despite high hopes noted above, Boeing engineers acknowledged many aerodynamic uncertainties in their efforts to design the aircraft that became the B-47:

> As we commenced to think in terms of higher speeds and wing loadings, it became immediately apparent that we were contemplating an airplane whose normal operation would be within the region in which the effects of compressibility are prevalent. Hardly any data existed at that time relative to either the character or magnitude of these effects, but indications were that drag would be adversely affected and stability and control characteristics could be dangerous. ... Performance predictions were necessarily based on extrapolations of skimpy available data, and on the assumption that the unknown compressibility effects could be minimized and safely handled, given sufficient time for aerodynamic development ... [It) was, and is, our opinion that the success of an airplane in the category of Model 424

depends largely on attainment of one characteristic speed. We therefore felt it unwise to rush into the construction phase of such an airplane without the reasonable assurance that this characteristic could be attained and safely utilized. [72]

Other technical problems included fabrication of components. For instance, the turbojet's nozzle, compressor blades, and turbine required material performance characteristics that were thought impossible until the early 1940s. [73] Components also had to be designed with very small manufacturing tolerances. [74] These problems required time for research, design, fabrication, and test activities. Table 1 shows the time required to develop early engines.

Recognizing the gaps in knowledge, NACA's Ames Laboratory and other government laboratories inaugurated new aviation research programs in which they investigated drag rise and other adverse effects arising from the shock waves caused by the motion of the airfoil, and the severity of these effects at higher speeds. This research revealed that a swept wing might serve in high subsonic and low supersonic speed ranges, but for higher supersonic speeds a thin low-aspect-ratio wing would be desirable.

Table 1
USAF Engine Development Time

Engine	J33	J35	J47	J57
First Proposal	5/1943	5/1943	2/1946	7/1947
First Run	1/1944	4/1944	6/1947	6/1949
Months to First Run	8	11	16	23
50-Hour Test	6/1945	2/1946	4/1948	8/1951
Months to 50-Hour Test	25	33	26	49
150-Hour Test	4/1947	12/1947	3/1949	11/1953
Months to 150-Hour Test	47	55	37	77
Source: Elapsed times between dates are approximate. Table drawn from Jacob Neufeld. Air Force Jet Engine Development, A Brief History, 2d ed. (Washington, D.C.: Office of Air Force History. 1990), vii.				

For the B-52, research progress was manifested in frequent design changes based on new knowledge about propulsion and aerodynamics. But new information about the difficulty of fielding an efficient turboprop engine and the potential of turbojet engines did not playa dominant role in design deliberations until Boeing's October 1948 Model 464-49. By this time, the XB-47 had been flying for 10 months. Pilots flying the aircraft liked it, and commented that "the plane still is doing much better than anyone had a right to expect." [75] And positive test results from the more powerful and fuel efficient turbojet engine, the J47, were in; the J47 passed its 50-hour test in April 1948, which validated the engine for experimental flight.

The B-52 design included a requirement for turbojet propulsion only when evidence had accrued about their potential performance. The following bullets outline the B-52's major design features leading to the jet-powered model.

• Model 462, a six-turboprop engine straight-wing airplane, was a first attempt to design an aircraft that could be improved, through incremental model changes, into a 5,000-mile radius heavy bomber.

• Model 464 represented a lighter, shorter range aircraft. It reflected the view that Model 462 was not optimal and that a 5,000-mile radius bomber would be too slow and far heavier than anticipated.

• Model 464-17 was the answer to questions about range and weight trade-offs. It concentrated on the atomic mission, sacrificed defensive armament, abandoned crew comfort, and reduced crew size to achieve desired range.

• Model 464-29 reflected improved aerodynamics and a new engine rather than a change in the basic concept of a long-range strategic bomber.

• Model 464-35 was stimulated by improved Soviet surface-to-air defenses that made the slow speed of the long-range 464-29 incompatible with the atomic mission. This model also took into account the improvements in aerial refueling as a way to achieve long range.

J47 Engine in XB-47

• Model 464-40's purpose was to explore the feasibility of turbojet engines in place of the turboprop power plants, but was never considered a serious design option.

• Model 464-49 employed swept wings and turbojet engines. This model grew partly out of 464-40, paper studies of the Pratt and Whitney J57 jet engine, the success of Boeing's B-47 medium bomber, approval of aerial refueling, and continuing difficulties in developing the T-35 turboprop engine.

Between 1946 and 1948, Air Force leaders emphasized turboprop development to avoid an error to which they were becoming sensitive: the failure to pursue a good idea. In the late 1940s, the evidence in favor of jet or turboprop power plants was ambiguous. [76] But believing turboprop power plants were the best near-term bet for military aircraft, about 60 percent of the Air Force engine development budget was devoted to turboprop study. This lack of diversity in the Air Force's investment strategy actually increased the probability of concentrating on an unpromising technology.

The choice of appropriate power plants involved many interrelated considerations, such as manufacturing technology, aerodynamics, engine design and performance. While technical evidence capable of deciding the issue was being generated by some companies (e.g., Pratt and Whitney, and GE), in the end, the findings of company jet research programs were overshadowed by other matters in deliberations about the B-52. Senior Air Staff officers were able to appreciate the value of jet propulsion for bombardment aircraft only after significant technical progress had been made.

Organizational Structure-Prewar and Wartime

A wide variety of events and conditions shaped the influence of the Air Force's organizational structure on the development of turbojet propulsion in long-range bombardment aircraft. In the interwar period, the evolution of aircraft engines was determined by the separate actions-and interaction-of Air Corps officers and senior decision makers, the NACA, and engine manufacturing firms. The center of the Air Corps' interwar activity in aviation technology was in the Power Plant Branch of the Engineering Section, located in Wright Field's Materiel Division. [77] The officers in the Power Plant Branch, mostly lieutenants and captains, had a clear goal: to create better engines, that is, generate more horsepower at less weight, minimize fuel consumption, reduce frontal area to reduce drag, and achieve maximum reliability at least cost in initial purchase and maintenance.

As noted earlier, these Air Corps engineers faced high "opportunity costs" for research on new engines. Ten years after World War I, more than 8,000 Liberty reciprocating engines were still available for use in military aircraft. The Power Plant Branch did not have funds to underwrite research on new engines for combat use. [78] Instead of funding new research, Power Plant officers drew up increasingly demanding specifications and then tested the resulting larger and more powerful engines industry developed. Air Corps Materiel Division "played an important role in the impressive achievements of industry" by testing the engines using rigorous standards for combat aircraft. In the mid-1930s, the Materiel Division insisted upon a 150-hour torque stand endurance test as well as before and after waterbrake or electric dynamometer tests to measure power output. After these tests, the engines were disassembled and scrutinized in the nation's best test facilities for signs of undue wear. [79]

The successes of engine firms and Wright Field monitors in advancing existing engine types concealed problems relating to technological planning and decision making in the Air Corps command structure. Between 1919 and 1926, the technical staff at McCook Field (after 1926, Wright Field) was heavily committed to "scientific investigation and fundamental research." Procurement was handled separately in Washington. D.C. [80] But in 1927, work on engine development was overtaken by Gresham's Law of Programming which observes that programmed behavior will drive out unprogrammed behavior. In this matter, aviation procurement was moved to the newly established Materiel Division to improve coordination between development and procurement functions. The unintended consequence of this reorganization was that administrative and testing duties associated with procurement assumed ever greater amounts of time and effort. In other words, the need to deal with immediate problems and administrative duties related to the reorganization drove out long-term developmental work on new technologies.

Procurement functions began to take a larger share of Wright Field's civilian and military personnel devoted to technical development-so that purely experimental research declined. In addition, firms were encouraged to absorb development costs because funds for experimentation were limited. Reimbursing the firms for development costs in subsequent production contracts discouraged innovations in engine design because of the long lag between initiation of trial-and-error engine research and subsequent production contracts. Firms engaging in incremental improvements of existing engine types could expect quicker and more certain returns on earlier development costs than firms attempting to develop new technologies for the long term. The engine firms' commitment to rapid incremental improvements of existing engine types kept the officers and equipment busy. The heavy testing "absorbed a great many engineering man-hours, further aggravating the prevailing scarcity of technically competent individuals." [81]

As an organization, the interwar Air Corps faced two obstacles in the task of advancing military technology. First, key decision makers were not adequately trained to make smart decisions about technological innovation. The education of Air Corps leaders "failed to develop adequate skills in objective analysis, in critical thinking, in separating fact from opinion, or in reaching conclusions only when warranted by verifiable evidence founded upon clearly recognized assumptions." [82] None of the officers who commanded the Air Corps had any engineering or scientific education above the undergraduate level. The officers who headed the Materiel Division during the interwar period also had no specialized engineering or scientific qualification. Further, some branch chiefs within the engineering sections did not even have engineering backgrounds. [83] The high-ranking materiel officers who attended the Army professional schools had no instruction on "the art, problems, and practices of technological planning and decision making." And the schools were not intellectually demanding. [84] Hence, as historian I. B. Holley Jr. argued, "indoctrination, rather than the cultivation of a capacity for critical thinking, was the dominant objective at the staff school." [85] The Army schools did not communicate the relationship between science and weapons development. And Air Corps officers, including General Arnold, were unreceptive to greater cooperation with universities on Air Corps research.

Second, the organizational structure of the Air Corps was unable to provide senior officers with information and analyses to identify technical challenges and opportunities they were facing in engine development (such as the practical limits on aircraft speed generated by reciprocating engines or-after 1935-the potential of turbines for high-speed flight). The shape of organizational hierarchy determines which organizational members handle particular problems and tasks, make comparisons and judgments, and approve or veto a proposal. Comparisons of information, options, and implementation proposals are critical elements in the policy- making process. Different organization structures can produce different policy outcomes through the way information and decisions flow up and down the organization. [86] In the Air Corps, technically talented officers were overwhelmed and over-occupied with testing duties. They also were not in administrative positions conducive to influencing the course of technological planning. The military leaders of Materiel Division, during the interwar period, had no specialized engineering or scientific training. Hence, those charged with providing analyses of aviation technology to senior leaders in Washington did not have the intellectual background to handle the increasingly technical character of the relationship between science and weapons development.

In sum, the structure of the interwar Air Corps organization hampered the ability of senior Air Corps officers to confront the difficult and complex questions of whether and how to advance military technology. First, research functions were marginalized in comparison to procurement and other policy matters considered by air leaders. Procurement dominated research, and scarce engineering talent was diverted from experimental work or testing. A brief comparison with the Navy's aviation community between 1919 and 1925 would be instructive. Although RAdm Moffett, through the Bureau of Aviation, controlled procurement-he could not dominate the analytic process that set military characteristics, designed new operational concepts, and integrated the two. The Navy's analytic process did not fall prey to the type of pitfalls found in the Air Corps because, although there was overlap of policy jurisdiction, the naval agencies were independent.

Second, the incentives built into the procurement policy favored incremental engineering development which, for all its substantial benefits, diverted attention away from long-range research and development. Third, a high percentage of officers assigned to headquarters positions overseeing research and development had minimal scientific or technical preparation. Thus, the Air Corps' leaders failed to understand the interaction between fundamental and applied aviation research. And adding to the problems associated with the Air Corps' structure was the fact that NACA's leaders protected their virtual monopoly on fundamental research in aeronautics by refusing to expand the range of engine research (until the outbreak of war) or to examine theoretical aspects of supersonic flight even after evidence of possible high value results (in 1935 Volta conference) became clearer.

Postwar Organizational Structure and Implementation

The post-war Air Force had its own barriers to the introduction of advanced technology. Some of these barriers carried over from the interwar period, for example, the poor technical preparation of officers in key R&D and acquisition positions. Organizational structure also affected acquisition decisions by presenting different trade-offs to officers in the field and at headquarters, by centralizing veto authority in headquarters staff but technical competence in the field, and by implicitly setting different rules of evidence for decisions, comparisons, and evaluations of information by field and headquarters personnel.

Officers working in the Bombardment Branch and Air Materiel Command laboratories (and supported by technically trained civilians) understood the difficult trade-offs inherent in building an advanced aircraft when technical knowledge was - growing rapidly and budgetary constraints limited freedom of choice. AMC officers were more insulated from the conflicts associated with unification of the services, demobilization, and disputes than were the officers at Air Force Headquarters. Hence, they could devote more time and attention to technical issues. In contrast to AMC officers, officers at Air Force Headquarters, who were primarily operators, were less able to understand the technical premises for their decisions. [87] Headquarters officers were more easily impressed by manufacturers' claims for unproven aircraft designs such as Northrop's Flying Wing.

The concerns of AMC officers focused on designing and constructing an effective aircraft. They balanced the aerodynamic and aeronautic trade-offs generated by changes in specific requirements through discussion of competing studies and bargaining and negotiating over design alternatives. This approach led to conflicts between the AMC and such headquarters senior leaders as Gen Carl A. "Tooey" Spaatz, who was concerned with the Air Force's status in relation to the Navy and Army. [88] The Air Staff conflict with AMC was fueled by the tendency of Air Staff officers to stipulate the Air Force position on technological matters. That is, Spaatz and his aides implicitly believed that their competence and experience in operational matters carried over to technological questions. But competence in the operational domain does not automatically entail competence in evaluating technology. In addition, the time and attention of senior Air Force leaders were dominated by political questions. Spaatz and his close advisers on the Air Staff, including Maj Gen Lauris Norstad and Lt Gen Ira C. Eaker, testified before Congress, lobbied, and dealt with the problems of demobilization, budgets, and unification.

Key to understanding the conflict between AMC and the assistant chiefs of Air Staff (AC/AS) over the design of the B-52 is the role of knowledge and analysis in decision making. The tasks and questions considered in AMC were relatively well defined technically. Experimental procedures, standards of proof, and a body of background knowledge structured AMC investigations. In contrast, the tasks and questions considered by the assistant chiefs at headquarters were ill defined. They had no off-the-shelf or proven way to win bureaucratic or political battles with the Navy over roles and missions. Such matters contain elements of symbolism, bargaining, and persuasion for which there is no unambiguous solution. Further, as indicated above, most of the officers in AC / AS staff positions were technically less well-trained than those at AMC and less able to comprehend the hardware issues they faced. Hence, the use of knowledge and analysis in AC / AS decision making was limited by negotiation over ill-defined issues.

Yet, Air Materiel Command constituted a useful "redundancy of calculation" for the assistant chiefs, and functioned as an element of a nascent multiorganizational system. [89] AMC staff often criticized the bases and assumptions of headquarters' B-52 decisions. A June 1947 memorandum from the Aircraft Laboratory argued that AC / AS staff (1) misunderstood the relation between military requirements and aircraft size, (2) misunderstood how difficult it would be to design an aircraft capable of 5,000-mile radius, (3) did not appreciate how well balanced the B-52 design was, and (4) misunderstood how technical setbacks should be expected but could be solved in later versions of the aircraft. [90] In July AMC's Maj Gen Laurence C. Craigie, arguing on the basis of technical studies conducted in the Engineering Division, suggested that AC / AS officers should refrain from proposing either the all-wing or delta wing as alternatives to the B-52. He emphasized, in response to misgivings about B-52 range, that design deficiencies could be rectified in the aircraft's life cycle. [91]

The Air Force hierarchy determined how existing technical knowledge would be employed to make acquisition decisions. The preoccupation of Air Staff officers with political matters, staffing, training, and budgets precluded attention to figuring out how to reduce technological uncertainty about propulsion. There was considerable disagreement about the ability of engines under development-which as yet existed only on paper-to meet the speed and range requirements. No power plant-propeller combination then available could cruise at speeds above 400 MPH at 35,000-feet altitude and carry a 10,000-pound payload for a 5,000-mile mission radius. The difficulties in achieving these goals led to a continual search for other means to accomplish the mission, leading to frequent alterations of military characteristics. Between November 1945 and December 1947, the heavy bomber's military characteristics were changed at least four times. Attempting to find a satisfactory airplane, Air Force Headquarters officers proposed technically premature, yet seductive, ideas for heavy bomber configuration-delta wing and all-wing designs.

In contrast, AMC's senior leaders emphasized increasing the amount of aerodynamic knowledge. Craigie, for example, recognized the likelihood of adopting a poor heavy bomber design in the context of the existing knowledge base, because the technical tasks of design and construction were formidable. Consistent with Craigie's approach, Col Henry E. "Pete" Warden, AMC's chief of Bombardment Branch, tried to fill gaps in knowledge about turbojet power plants. He had argued in December 1947 to a doubting audience at Air Force Headquarters that turbojets were more feasible for heavy bomber propulsion than turboprops. He also urged Pratt and Whitney engineers and managers, without the knowledge of Air Force Headquarters officers, to work on a turbine capable of either combining with propellers or becoming a pure turbojet.

Warden's evaluation of turboprop versus turbojet propulsion was based partly on faith that later turbojet models would demonstrate improved performance. This faith was based on wartime experience concerning the steady performance improvements propulsion engineers were able to wring out of the new jet engines. Warden's faith also was based on a rigorous analysis of the speed limitations, increasing mechanical complexity, long development times, and declining marginal increases in performance of turboprop power plants. [92] The issue here is not whether Warden's evaluation of the potential of turbojet propulsion was correct, it is the character of premises used in decision making. Warden's arguments were consciously based upon better formulated and justified technical premises than were the arguments supporting the all-wing or delta aircraft made by Air Force Headquarters officers. Warden's conclusions were correct for the right reasons.

Not all headquarters discussions regarding the heavy bomber program relied upon faulty technical premises. [93] Nevertheless, most headquarters officers did not have the time, energy, or training to focus on technical premises for design decisions. Their attention was focused on the problems of justifying and securing independence for the Air Force and defending the Air Force strategic role and mission in national defense.

The Political Environment of Postwar Bomber Acquisition

The tremendous growth of the Air Corps during World War II created a plethora of offices with overlapping concerns-the organizational conditions for the establishment of a multiorganizational system that ultimately influenced the postwar formulation of operational requirements for jet-propelled strategic bombers. Peacetime, however, brought large cuts in personnel, budget, and orders to aircraft manufacturers and-against the background of great uncertainty and strife in the defense establishment-many doubts surfaced during the first 13 months of the new heavy bomber program. The first bomber configuration, Model 462, was accepted in mid-1946. Within three months, this version was subjected to much criticism by the Air Staff. Boeing proposed an entirely different configuration, Model 464, to answer Air Staff doubts. Over the course of several years, the profusion of offices having overlapping functions regarding the development of new aircraft promoted a useful pattern of proposal, criticism, and change.

In the context of peacetime political battles, AMC B-52 program managers pursued the difficult task of applying growing aerodynamic knowledge to their evaluations of B-52 design. Time and again between 1946 and 1948, Air Staff officers rejected Boeing bomber proposals because of projected range and speed shortfalls. Air Staff officers in charge of formulating military requirements (in consultation

48

with AMC Bombardment Branch officers) responded to evaluations of inadequate aircraft performance by altering military characteristics-required size and range were reduced in return for higher speed.

The year 1948 began under a dark cloud for AMC's B-52 program managers. Air Staff officers succeeded in canceling, not simply Boeing Model 464-29, but the entire Boeing heavy bomber program due to doubts about the B-52's ability to achieve the required range and speed. Some Air Staff officers preferred the Northrop YB-49 turbojet powered all-wing aircraft over Boeing's conventional B-52 design: others favored opening a new competition for a heavy bomber.

Rapid progress on an acceptable heavy bomber design then was stalled in the early months of 1948 while Boeing president William M. Allen and AMC officers lobbied Air Force Secretary Symington and headquarters officers to reinstate Boeing's contract. During this period, despite the cancellation, Boeing and AMC engineers continued their discussions and research on heavy bomber design. This activity led to the Boeing proposal for Model 464-35 after Symington and, Air Force Undersecretary Arthur S. Barrows reestablished the Boeing contract. While several compromises in military characteristics were made to give Model 464-35 a better chance of meeting Air Force needs (e.g., reduced required range), technical shortcomings in the fire control system, landing gear, engine nacelle design, and aircraft configuration still made achievement of military characteristics dubious.

Senior Air Force leaders had not set out to design and build a swept-wing long-range jet bomber. Yet a turbojet swept-wing aircraft design (Model 464-49) was proposed in October 1948 as a way to avoid the many difficult technical problems associated with turboprop power plants. The lack of coherence in this development process was, in fact, typical, according to Michael E. Brown. Surveying 15 major post-World War II strategic bombardment programs, Brown argued that Air Force leaders did not understand what they were doing when they initiated many of their bomber programs.

> In many cases emerging operational threats were far from clear when they set performance requirements for new systems ...the technological possibilities for weapon system development were rarely well understood, because the Air Force decision makers generally failed to make a thorough assessment of the technological horizon before they launched new ventures. They routinely compounded the unknowns their programs faced by setting performance requirements far beyond the state of the art. [94]

The B-52's development, indeed, featured widespread ambiguity in terms of objectives (the requirements changed), technology (technological options changed), and participants (senior officials were reassigned). There was no direct and causal link between the initial conception of the aircraft and the final configuration. The character of the threat remained unclear.

The meaning and implications of this history are critical for national security and acquisition officials. Many would argue that the proper way to deal with incoherence in military development programs is to streamline, increase vertical coordination, and to enforce coherence in policy and decision making. Experience has shown, however, that the incoherence of decision making is unavoidable in individual organizations dealing with new technology or operational concepts, and that efforts to impose coherence are counterproductive. A more effective adjustment to ambiguities caused by changing objectives, technology, and participants is to establish a multiorganizational system capable of proposing, reviewing, comparing, and evaluating new ideas.

The formal decision process for managing development programs "rationally" resembles the structure of most American military operational or combat decisions: using a peaked hierarchy and a tightly coupled, linear flow of communication. Yet, the following chapter will show that the smart decisions that led to acceptance of the Boeing Model 464-49 design (a swept-wing aircraft powered by eight Pratt and Whitney J57 turbojets) were actually the product of an informal multiorganizational system.

Officers assigned to technical positions at AMC struggled with officers at headquarters over bomber design. The resolution to these disagreements emerged from criticism of the Boeing proposals by various Air Force offices and resulting discussions between the Boeing partisans and opponents. The process generating the criticism was not set up by top Air Force officials when they began to organize the post-war Air Force. In fact, the relations between AMC and headquarters constituted a fairly effective, if unplanned, way to utilize knowledge and analysis in decision making. While the formal organizational system maintained the primacy of rank in command, AMC was sufficiently independent of the Air Staff to permit the authority of knowledge and parallel analysis of, policy to have a great impact on the decisions made.

Notes

1. Robert Schlaifer, Development of Aircraft Engines (Boston: Graduate School of Business Administration, Harvard University, 1950), 485.

2. Jet propulsion was not the only significant interwar technology that the Army aviation community ignored. The controllable-pitch propeller would have been abandoned entirely if it had depended on military support. Instead. it was discovered that Boeing 247 transports could not take off from high-altitude airfields in the Rockies carrying a reasonable payload without controllable-pitch propellers. The resulting orders led to the first economic manufacture and future development of controllable pitch propellers. Schlaifer, 57.

3. The National Defense Act of 1920 created an Army of the United States whose core issue was manpower mobilization. See Allan R. Millet and Peter Maslowski, For the Common Defense: A Military History of the United States of America (New York: Free Press. 1984), 365-68.

4. James G. March, with the assistance of Chip Heath, A Primer on Decision Making: How Decisions Happen (New York: Free Press, 1994), 248.

5. Martin Landau. "On Multiorganizational Systems in Public Administration," Journal of Public Administration Research and Theory I, no. 1 (1991): 9.

6. See also Norman Friedman. Thomas C. Hone, and Mark D. Mandeles. The Introduction of Carrier Aviation into the U.S. Navy and Royal Navy: Military-Technical Revolutions, Organizations, and the Problem of Decision (Washington, D.C.: OSD/NA, 1994).

7. The Advanced Development Objectives for a turbojet-powered fighter aircraft were released in July 1941. The first American jet fighter, Bell's XP-59A Airacomet. was ordered in September 1941. This aircraft, propelled by a General Electric engine (designated the I-A), was based on the Whittle engine.

8. Schlaifer, 440.

9. I. B. Holley Jr., "Jet Lag in the Army Air Corps," in Military Planning in the Twentieth Century: Proceedings of the Eleventh Military History Symposium, 10-12 October 1984, edited by Harry R Borowski (Washington, D.C.: GPO, 1986), 124-25.

10. The great numbers of war-surplus Liberty engines, many unused, for sale at a fraction of cost also stifled research on advanced reciprocating engines. It was only in 1934 that the Army halted the use of Liberty engines to power Army aircraft because Representative Fiorello H. LaGuardia, recognizing the faults and obsolescence of the engines, inserted a rider into the Appropriations Act mandating an end to the use of Liberty engines in Army aircraft. Schlaifer, 160: Holley, 133.

11. Schlaifer, 445.

12. Holley, 125-26.

13. Edward W. Constant II, The Origins of the Turbojet Revolution (Baltimore: Johns Hopkins University Press, 1980), 154-56.

14. Constant, 156-59: Holley, 126. 15. Schlaifer, 87.

16. Constant, 180-92; Holley. 127. 17. Holley. 127.

18. Alfred Goldberg, "The Quest for Better Weapons," in The Army Air Forces in World War II, vol. 6, Men and Planes, ed. Wesley Frank Craven and James Lea Cate (Washington, D.C.: GPO. 1983), 247.

19. Schlaifer, 446-48.

20. Goldberg, 247.

21. For a description of NACA's neglect of engine research during the interwar period, see Alex Roland. Model Research: The National Advisory Committee for Aeronautics 1915-1958 (Washington, D.C.: NASA, 1985), 187-94.

22. The committee also considered rocket propulsion.

23. Durand was an energetic chairman and jet propulsion was considered very seriously. In September 1941, the committee recommended that contracts to build turbojet engines be given to each of three companies, Schlaifer, 459-60.

24. Schlaifer, 458-61. 25. Ibid., 461.

26. Jacob Neufeld, "Air Force Jet Engine Development, A Brief History," 2d ed. (Washington, D.C.: Office of Air Force History, 1990), 2.

27. Schlaifer, 461; Goldberg, 248.

28. There was very little public information about jet propulsion. One of the few reports concerns an Italian jet designed by Campini. Alexander Klemin, "Jet Propulsion," Scientific American 165, no. 5 (May 1942): 251.

29. The restrictions continued until the summer of 1943. Neufeld, 3.

30. Memorandum, R. F. Bradley to E. C. Wells, 15 August 1944, Boeing archives. Cited in Michael E. Brown, Flying Blind: The Politics of the U.S. Strategic Bomber Program (Ithaca: Cornell University Press, 1992), 74.

31. Schlaifer, 490. 32. Ibid., 468.

33. Ibid., 321; Goldberg, 246; Holley, 123; Roland, 189.

34. The expenditures at Hinkle for a comparable period of time were only slightly higher, and Hinkle produced a flying jet-powered aircraft, Schlaifer, 483-84.

35. Holley, 130-31.

36. See Robert T. Finney, History of the Air Corps Tactical School 1920-1940 (Washington, D.C.: Center for Air Force History, 1992), 55-78.

37. Herman Wolk, Planning and Organizing the Postwar Air Force 1943-1947 (Washington, D.C.: GPO, 1984). 19-20.

38. John F. Shiner, Foulois and the U.S. Army Air Corps (Washington, D.C.: GPO, 1983).

39. While Holley criticizes Arnold's technical education and grasp of " technology, Arnold was able to infer military potential from some research results, e.g., rocket propulsion. Arnold's insight was strengthened by personal tutoring from von Kármán on a number of topics in aeronautics. See Michael H. Gorn, The Universal Man. Theodore von Kármán's Life in Aeronautics (Washington, D.C.: Smithsonian Institution Press, 1992), 83-85.

40. Col J. S. Holtoner, chief, Aircraft. Branch Office, AC/AS-4, Research and Engineering Division, memorandum to Generals Carroll and G. Gardner, subject: Hanson Baldwin's "Problems of the B-36," from the New York Times, Thursday, 10 July 1947, 14 July 1947, Library of Congress, Hoyt S. Vandenberg Collection, box 34, "AC/AS-4."

41. Maj Gen Grandison Gardner, acting AC/AS-4, memorandum to Gen Hoyt S. Vandenberg, vice chief of staff, subject: Problem of the B-36, 14 July 1947, library of Congress. Hoyt S. Vandenberg Collection, box 34, "AC/AS-4."

42. Holtoner.

43. Goldberg, "The Worldwide Air Force," 224. See Elliott Vanveltner Converse III, "United States Plans for a Postwar Overseas Military Base System, 1942-1948" (PhD diss., Princeton University, 1984), 255-56; Melvyn P. Leffler, "The American Conception of National Security and the Beginnings of the Cold War, 1945-1948," American Historical Review. April 1984, 349, 352-53.

44. John E. Steiner, "Jet Aviation Development: A Company Perspective," in The Jet Age: Forty Years of Jet Aviation. ed. Walter E. Boyne and Donald S. Lopez (Washington, D.C.: Smithsonian Institution Press, 1979), 146.

45. Marcelle Size Knaack, Encyclopedic of U.S. Air Force Aircraft and Missile Systems, vol. II, Post-World War II Bombers 1945-1973 (Washington, D.C.: GPO, 1988), 101.

46. Steiner, 146.

47. Maj Gen Curtis E. LeMay, DC/AS for R&D to Lt Gen Nathan F. Twining, commander, AMC, letter, 15 May 1947.

48. David Alan Rosenberg, "U.S. Nuclear Stockpile, 1945 to 1950," The Bulletin of the Atomic Scientists, May 1982, 27.

49. Bowen, A History of the Air Force Atomic Energy Program. 57. 50. Memorandum, Putt to Spaatz, 26 January 1948. Air Staff Summary Sheet, prepared by Gen Lauris Norstad, subject: Weight Reduction and Simplification of B-52 Airplane, 6 October 1948.

51. Brig Gen S. R. Brentnall, deputy director, Research and Development, AMC, memorandum to Gen Hoyt S. Vandenberg, chief of staff of the Air Force (CSAF), attn.: DCS/M. subject: Weight Reduction and Simplification of the B-52 Airplane," 20 January 1949: A. G. Carlsen, Boeing chief project engineer to M. P. Crews, letter, subject: Contract W33-038 ac-15065-Model XB-52 Performance Data, 21 January 1949.

52. John L. Gaddis, Strategies of Containment: A Critical Appraisal of Postwar American National Security Policy (New York: Oxford University Press, 1982). 57.

53. Paul Y. Hammond, "Super Carriers and the B-36 Bombers: Appropriations, Strategy and Politics," in American Civil-Military Relations, ed. Harold Stein (Birmingham, Ala.: University of Alabama Press, 1963).

54. Gordon Swanborough and Peter M. Bowers, United States Navy Aircraft Since 1911, 2d ed. (Annapolis, Md.: Naval Institute Press, 1976), 186.

55. Ibid.. 186-88. 56. Constant, 249.

57. Ibid., 262.

58. Ibid., 263.

59. Schlaifer, 321.

60. Neufeld, ii-iii.

61. General Electric Company, Eight Decades of Progress: A Heritage of Aircraft Turbine Technology (Cincinnati: The Hennegan Co.. 1990), 38.

62. Ibid., 42-45. 63. Neufeld, 3-6.

64. Edwin P. Hartman. Adventures in Research: A History of Ames Research Center 1940-1965 (Washington, D.C.: NASA, 1970), 119.

65. Steiner, 142.

66. Gen F. O. Carroll, chief, Engineering Division, Materiel Command to Development Engineering Branch, Material Division, AAF, letter, subject: XB-47 and B-47 Aircraft. RG 18, AAF, NARS, 8 November 1944, cited by Brown, 74.

67. Hartman, 121.

68. Theodore von Kármán, "Some Significant Developments in Aerodynamics Since 1946," Journal of the Aero/Space Sciences, March 1959, 129-44, 154.

69. Hartman, 155.

70. Steiner, 143-46. 71. Hartman. 123.

72. W. E. Beall, memorandum to C. L. Egtvedt, 19 April 1945, Boeing Archives. Cited in Brown, 76.

73. John A. Miller, Men and Volts at War: The Story of General Electric in World War II (New York: McGraw-Hill Book Co., Inc., 1947), 80.

74. Constant, 258.

75. Knaack, Post-World War II Bombers, 106-7.

76. Thomas A. Marschak, "The Role of Project Histories in the Study of R&D," in Econometrics and Operations Research, vol. 8, Strategy for R&D: Studies in the Microeconomics of Development, ed. Thomas A. Marschak, Thomas K. Glennan Jr., and Robert Summers (New York: Springer-Verlag, 1967).

77. Holley, 132.

78. Ibid., 133.

79. Ibid., 134.

80. Ibid.. 135.

81. Ibid., 136.

82. Ibid., 141.

83. Ibid., 138.

84. Ibid., 139.

85. Ibid., 140.

86. See Jonathan B. Bendor and Thomas H. Hammond, "Rethinking Allison's Models," American Political Science Review 86, no. 2 (June 1992): 317.

87. See discussions of technical training in Holley, 137-42: Alan L. Gropman, "Air Force Planning and the Technology Development Planning Process in the Post-World War II Air Force-the First Decade (1945-1955)," in Military Planning in the Twentieth Century, ed. Harry R. Borowski (Washington, D.C.: GPO, 1986), 161-63.

88. Officers at Air Force Headquarters worked in offices of Assistant Chief of Air Staff/Operations and Training (AC/AS-3), Assistant Chief of Air Staff/Materiel (AC/AS-4), and Deputy Chief of Air Staff for Research and Development (DC/AS for R&D).

89. Martin Landau, "Redundancy, Rationality, and the Problem of Duplication and Overlap," Public Administration Review 29, no. 4 (July/August 1969): 346-58: Mark D. Mandeles, "The Air Force's Management of R&D: Redundancy in the B-52 and B-70 Development Programs" (PhD diss., Indiana University, 1985), chap. 1.

90. J. A. Boykin, Aircraft Projects Section, AMC, memorandum report, subject: "Analysis of the XB-52 Project, 23 June 1947.

91. Maj Gen Laurence C. Craigie, chief, Engineering Division, AMC to Gen Carl A. Spaatz, commanding general AAF; attn.: AC/AS-4, letter, subject: XB-52 Airplane, 11 July 1947.

92. Col Heruy E. "Pete" Warden, USAF, Retired, interviewed by author on 26 August, 27 August, and 9 September 1984.

93. Prompted by comments from headquarters staff, AMC officers noted that a major drawback of the Model 464-17 was that airfields then in use could not support a 400,000-pound bomber. Col John G. Moore, Aircraft Laboratory, Engineering Division, AMC, memorandum to Col George Smith, Bombardment Branch, Aircraft Projects Section, Engineering Division, AMC, subject: Reply to Routing and Record Sheet. 13 January 1947, re: "Landing Gear Requirement for Very Large Airplane," 18 February 1947.

94. Brown, x.

Chapter 5
The Introduction of Jet Propulsion into the B-52

The XB-52 is designed to supersede the B-36 as a long range strategic bomber. Its combination of aerodynamic refinement and turbo-propeller engines are the only presently known means of achieving characteristics of both long range and high speed-large improvements in this class of aircraft will come with radical developments which will require completely new airframe developments-[u]nless supersonic propellers become a reality, future [large bombers] will be powered by turbojet engines. However, neither of these developments are sufficiently near at hand that the turboprop step can be eliminated.

-Lt Gen Howard Arnold Craig

The opening quotation from a letter written by Lt Gen Howard A. Craig, the deputy chief of staff for materiel, illustrates the risks and pitfalls of decision making under uncertainty, ambiguity, and imperfect information in a large and complex organization. Craig, the senior Air Staff officer concerned with R&D, wrote with certainty about state-of- the-art aerodynamic and propulsion technology. His prediction was overturned in less than two weeks. As we explore how to organize to foster innovation, we must examine the characteristics of organizations and institutions that permit individuals to make better decisions consistently. How can we recognize and duplicate effective decision processes that both create new systems and develop the operational concepts to use them?

The American tendency has been to argue that there is a single "right method" of organizing. But, there is good evidence that performance of that right method degenerates over time into "looking smart rather than being smart," mistaking erudition for wisdom, and emphasizing style over substance. The only truly right method may be one that permits and encourages interaction, access, and criticism, but is silent on procedural specifics. Access to decision makers, interaction of people, and criticism of ideas, plans, and programs are essential to initiating the right kinds of changes. Effective innovation emerges from institutional rules and organizational arrangements permitting disagreement, competition, and communication.

This chapter presents a detailed chronology of the development process leading to the turbojet-powered swept- wing Model 464-49 which, with more modifications, eventually became the B-52. The level of detail presented here-the various options explored, arguments, disputes, and disagreements-is important to provide an accurate picture of the developmental policy process and to dispel unrealistic assumptions or prescriptions about the decision process. Understanding the full complexity of the interactions among people, organizations, and policies as the B-52 developed shows that neither a single unchanging "vision," nor order and harmony are necessary for a successful program.

The B-52's development is a near-ideal case of decision making under uncertainly, ambiguity, and imperfect information: problems faced by decision makers were ambiguous; technology and goals were unclear; problems and solutions were linked by availability (rather than foresight and prescience); some information collected by participants had only a tenuous substantive relationship to the problem under scrutiny; key personnel moved in and out of involvement in the program and did not have the time, experience, or understanding to address the problems before them; and decision makers did not understand the relationship between their decision style and outcomes. How then did this development process produce an aircraft of such high design quality that some models will remain in service into the next century-more than 50 years after the first models were first built?

Three aspects of the B-52's history are striking because they challenge conventional wisdom about "rationally and efficiently" managed or guided innovation. First, the Air Force personnel working on the B-52 did not end up with the aircraft they assumed they would get when the program began. The bomber's final swept-wing configuration, including the adoption of turbojet power plants, was very different from the initial conception of designers and Air Force leaders. [1]

Second, the development process did not conform to idealized features of a rationally organized program. While a rationally organized program has clear goals, adequate information, and well-organized and attentive leadership, the B-52 development process exhibited substantial disagreement over, and revision of, requirements or goals. Information about aerodynamic and propulsion technology was ambiguous, imperfect, and changing; the offices possessing the requisite technical expertise did not have corresponding decision authority; offices having less technical expertise prevented closure on less satisfactory early designs; and political concerns frequently focused senior leaders' attention away from the B-52 as they sought to create an independent Air Force and safeguard its status and missions.

Third, the discussions, analyses, and memoranda transmitted between Air Materiel Command and Air Staff officers forestalled premature closure on a particular design and allowed participants to learn as they went along. The competitive interaction of personnel from the AMC; Air Staff; airframe, engine, and propeller firms; and RAND spurred this learning and the continuous introduction of new knowledge into the design. As we shall see, this interaction of independent firms and Air Force offices performed, in effect, as a multi organizational system, identifying, preventing, and correcting errors that would have been overlooked in an ideal, streamlined, "rational" organization.

To explain the B-52's design process and how it succeeded, this chapter examines (1) the political environment for acquisition decisions, (2) uncertainties and ambiguities regarding technological goals and options, (3) the relationship of the Air Staff to AMC, and (4) the impact of uncertainty and imperfect knowledge on the decision process. Table 2 summarizes the development process with respect to requirements changes.

1945	• Operating radius of 5000 miles. • Speed of 300 MPH at 35,000 feet altitude. • Minimum payload of one 10,000-pound, Grand Slam configuration bomb, maximum payload of 80,000 pounds in various sizes of general purpose bombs. • Minimum crew of five, an undetermined number of 20-millimeter cannon operators of offensive and defensive armament and a six-man relief crew. • Armor protection for crew, fuel, engines, and other vital components consistent with weight and performance. Reliability, ease of maintenance, reduction in fire hazards, good visibility, quick change features, and simplicity of design.	miles. Speed 440 MPH over target. Payload of 10,000 pounds. Maximum gross weight of 400,000 pounds. Tapered straight wing. Powered by six Wright Aeronautical Corporation XT-35 gas turbine engines combined with propellers.	462.
September - October 1946			Air Staff: Boeing Model 462 six-turboprop engine design does not meet range requirements.
26 November 1946		Model 464: Maximum gross weight of 230,000-pounds, powered by four turboprop engines.	AMC: recommends accepting Model 464.
27 November	DC/AS for R&D and other Air Staff offices propose: • Operating radius of 5,000 miles and. reserve of 2,000 miles with one "Fat Man" Mk III bomb. • Atomic bomb mission only; maximum payload of 20-30,000 pounds. Tail armament only; neither all-around protection nor parasite fighter. • Minimum crew size. • Procure only one wing.		Air Staff: Boeing should initiate a study to design a bomber to meet long range atomic attack requirements
January 1947		Boing presents two aircraft configurations to meet the November requirements. Model 464-16: a tapered straight wing airplane powered by four turboprop engines (T-35 gas turbines). • Range of 13,800 miles. • Atomic payload only of 10,000-pound bomb. • Cruising speed of 420 MPH, high speed at target of 440 MPH. • No armor or defensive fire. • Maximum gross weight of 400,000 pounds. Model 464-17: similar to 464-16, but could carry a 90,000-pound conventional, high-explosive bomb load.	
March 1947		Boeing proposes several aerodynamic improvements Model 464-17, which leads to Model 464-29: Cruising speed of 455 MPH. Range of about 9,000 miles. Maximum gross weight of 400,000 pounds.	Air Staff accepts Model 464-17, but expresses concerns for fuel protection.
May 1947			DC/AS for R&D reiterates atomic mission: Range at least 10,000 miles. Payload of 10,000 pounds. High cruise speed. Model 464-29 appears to be good choice.

Date			
29 June 1947	Air Staff proposes: • Operating radius of 5,000 miles with one 10,000-pound bomb. • Average speed of 420 MPH. • Tactical operating altitude of 35,000 feet • Service ceiling of 40,000 feet.	Matches Boeing Model F-4-29	
July 1947			DC/AS for R&D, LeMay suggests AMC study other ways to accomplish the mission assigned to the B-52, e.g.: • One-way flight to the target (the aircraft returns to friendly territory outside the United States. • Planned ditching areas. • Use of subsonic pilotless aircraft.
October 1947	Range of 8,000 miles (approximately, 3,000-mile combat radius). • Cruising speed of 550 MPH and 550+ MPH high speed over defended area. • Droppable landing gear • Full purging and self-sealing tanks for fuel. • Minimum crew (pilot, relief pilot, navigator bombardier, weaponeer, gunner). .Air refuelability		Air Staffs Committee on Long Range Bombardment recommends amending heavy bomber characteristics and asks AMC's Aircraft Laboratory to examine new military characteristics
mid-November 1947			Air Staff recommends change order of contract to Long Range Bombardment Committee's requirements.
mid-November 1947			Air Staff recommends canceling Model 464-29, and developing an airplane capable of: • Cruising speed of 500 MPH. • Range of 8,000 miles. • Maximum gross weight of approximately 300,000 pounds.
28 November 1947	SAC proposes: • Speed of 520 MPH over 4,000 miles of enemy territory. • Range of 8,000 miles. • Tail armament only. • Payload of 10,000-pound bomb. • Maximum gross weight of 280,000 pounds. • Turboprop, rather than turbojet, propulsion (due to jet's high fuel consumption).		DCS/M proposes: • Cruising speed of 500 MPH, • Absorb all recent aeronautical advances, • Cancel Model 464-29, • Reopen competition. [No Air Force action on proposal.]
1 December 1947	The Perkins Committee—an ad hoc advisory committee for the Air Staff recommends: • Cancel Model 464-29. • Range of 8,000 miles. • Cruising speed of 500 MPH. • Maximum gross weight less than 300,000 pounds.		
2 December 1947			Air Staff representatives propose accepting requirements of maximum gross weight about 300,000 pounds and 8,000-mile range at 500 MPH cruising speed. AMC disputes range projections for turboprop aircraft and concludes that

mid-Decem* of 1947			Air Staff cancels Boeing's B-52 contract and reopens design competition.
mid-February 1948		Boeing proposes Model 464-35: • Maximum gross weight of 300,000 pounds. • Range of about 8,000 miles. • Speed of 500 MPH over 4,000 miles of enemy territory. (This proposal matched November 1947 requirements.)	
3 March 1948	New requirements circulated on Air Staff: • Range of about 8,000 miles. • Tactical operating altitude of 40,000 feet, with 45,000 feet desired. • Speed of 500+ MPH, with 550 MPH desired.		
15 June 1948			Air Staff concludes Boeing Model 464-35 is too similar to Model 464 and will not match the March 1948 requirements.
28 July 1948		Model 464-40: • Eight turbojet power plants. • Maximum gross weight of 280,000 pounds. • Range, with 15,000-pound payload and high-speed run over 4,000 miles of enemy territory, of 6,750 miles. • High speed at 35,000 feet altitude in target range of 536 MPH. • Cruising speed of 483 MPH. • Service ceiling at target weight of 45,200 pounds.	AMC requests Boeing study turbojet propulsion.
October 1948			DCS/M asserts: Model 464-35 represents "the only presently known means of achieving characteristics of both long range and high speed… aerodynamic and structural margins… have been stretched farther than hs been the case in past designs… Growth through a series of models similar to that of the B-29 is not visualized." Range improvement, if any, would come in improvements in fuel economy and by aerial refueling.
mid-October 1948		Model 464-49: • Swept-wing, eight turbojet power plants. • Maximum gross weight of 330,000 pounds. • Payload of 10,000 pounds. • High speed at target of 560 MPH. • Target altitude of 49,400 feet. • Range of 6,750 miles.	AMC urges Boeing to develop swept-wing turbojet design.
late-1948 to 1951	Air Staff favorably impressed with characteristics projected for 464-49; near-cancellation of B-52 program in 1949-1950; production decision made in January 1951.		

The Emerging Postwar Security Environment

Shortly after the end of World War II, Air Force leaders, looking ahead to possible conflict with the USSR, sought to remedy the perceived range and payload deficiencies of available aircraft, that is, the B-29. On 23 November 1945, the Air Staff issued the first post-World War II series of military characteristics for new heavy bombers. These characteristics specified the level of performance-the requirements to meet the military threat-expected for the next generation of bombers, including the following:

• An operating radius of 5,000 miles (the operating radius is the distance between an aircraft's takeoff point and ultimate point of safe return). [2]

• A speed of 300 miles per hour at 35,000-feet altitude.

• A minimum payload of one 10,000-pound, Grand Slam configuration bomb (yielding the equivalent of 80,000 pounds of explosives); a maximum payload of 80,000 pounds in various sizes of general purpose, conventional high -explosive bombs.

• A minimum crew of five, an undetermined number of 20-millimeter cannon operators for offensive and defensive armament, and a six-man relief crew.

• Armor protection for crew, fuel, engines, and other vital components consistent with weight and performance.

Reliability, ease of maintenance, reduction in fire hazards, good visibility, quick change features, and simplicity of design. [3]

Due to the secrecy enveloping the atomic bomb, these requirements only implicitly included the capability to carry atomic weapons by specifying a minimum payload of 10,000 pounds: the approximate weight of the "Fat Man" and "Little Boy" bombs.

As noted in the previous chapter, Air Force leaders emerged from World War II convinced that the strategic air mission had justified its independence from the Army. [4] They had argued that the way to win the European war was through strategic bombardment, rather than invasion of the continent by ground forces. Moreover, supporting troops in contact with the enemy detracted from the Air Force's ability to carry out the strategic air offensive rationally. Hence, Air Force officers wanted to specify the number, type, and importance of factors to be considered in air plans without being subordinate to ground commanders. Successive Army chiefs of staff, Gen George C. Marshall and Gen Dwight D. Eisenhower, were swayed by such arguments and supported a separate Air Force, coequal with the Army and Navy. [5] Postwar military planners, including Marshall, looked to airpower rather than a large standing army to avoid future wars. [6] If war did occur, postwar planners reasoned, it would be fought first in the air, since surface forces could not operate successfully without air superiority. [7]

The command reorganization of the Air Force after the war "outlined by Eisenhower and [Gen Carl A. "Tooey"] Spaatz was keyed to the establishment of the Strategic Air Command (SAC) which was visualized as a long-range striking force equipped with atomic capable B-29s and B-36s." [8] Formed in March 1946, along with Tactical Air Command and Air Defense Command, SAC was to be the core of the Air Force effort to rebuild its combat readiness. The plans for SAC were based on scientific reports that fissionable materials were very scarce and expensive. Air Force leaders planned that SAC would deliver both nuclear and conventional bombs. However, in 1946, SAC's striking power was very weak. The B-29s of the "atomic capable" 58th Bomb Wing were not a credible deterrent and only one unit, the 509th Composite Group, actually possessed aircraft equipped to carry atomic bombs. Thus, SAC's ability to perform a sustained, long-range nuclear operation was negligible. Also, SAC was equipped inadequately for conventional operations; its B-29s could not attack Soviet targets from the continental United States. The United States had no adequate forward bases in Europe to attack European Russia (Japanese bases were too far away from those targets), and SAC had not yet developed effective aerial refueling to extend the range of its aircraft.

Indeed, due to the rapid demobilization with the end of World War II, the Air Force was unable to respond to its first post-war crisis in August 1946, when two American C-47 Skytrain transports were shot down over Yugoslavia. The State Department proposed an immediate and aggressive use of airpower over that country. Maj Gen Lauris Norstad (AC/AS-5, Plans) had to respond that the Air Force was too weak to risk war. [9] This recognition stirred greater concern in Air Force leaders to assure intercontinental range (without sacrificing payload) in their aircraft.

Air Force planning officers did not concentrate solely on responding to military threats. Quiet periods allowed consideration of other issues. Indeed, political problems at home demanded Air Force leaders' attention as they dealt with post-war budget reductions and demobilization. The number of Air Force personnel dropped from a wartime high of 2,411,294 in March 1944 to 353,143 in January 1948.

During a period of especially rapid personnel reduction in 1946, Commanding General Carl A. Spaatz worried whether he could "salvage something" of the Air Force. World War II experience showed that an operational air force could not be built quickly. [10] Air Force leaders responded to the demobilization by being the most persistent and vocal advocates of constant military readiness. They argued against the old mobilization idea of waiting for a declaration of war before preparing the economy and military to fight. Such a policy before World War II had resulted in the tardy development of a plan for all-out mobilization of America's resources. A mobilization plan achieved definition only a few days prior to the Japanese attack on Pearl Harbor. It took two more years before administrative methods were developed to manage American resources to meet wartime operational needs. [11] These arguments, of course, were part and parcel, of the efforts to create an independent Air Force.

Demobilization of the Air Force, however, proceeded rapidly over the objections of Air Force leaders. In February 1945 there were 243 Air Force groups worldwide. Assistant Secretary of War for Air, W. Stuart Symington (with the help of Spaatz), argued for a postwar force of 70 groups. Yet, by December 1946, there were 55 groups, only two of which were combat ready. [12]

As demobilization proceeded, military planners disagreed over reductions in the Air Force budget and the number of wings and new weapon systems to build for combat echelons, especially SAC. Even greater clashes emerged from the struggle to unify the armed services and to define roles and missions for each service. This conflict led to an Air Force independent from the Army, "but tied," in Trevor Gardner's words, "in triple harness to the peculiar concept of 'balanced forces' and of policies geared to service, rather than to national roles and missions." [13]

The efforts between 1945 and 1947 to unify the services and to create an independent Air Force were divisive. The intense bureaucratic conflict between the Air Force and Navy was a feature of this period's security environment because the competition between the services influenced the allocation of budget authority and the way each service would structure itself to fight. For several years after creation of the National Military Establishment (later called the Department of Defense), the Navy was the chief bureaucratic opponent of the Air Force as both vied for the main role and strategic mission in any future war.

Part of the Navy's difficulties with the Air Force stemmed from the internal Navy struggle to redefine its doctrine and role in the new national security arrangements. In February 1946 the Joint Strategic Survey Committee, Army Chief of Staff Dwight D. Eisenhower, and General Spaatz agreed to define service missions in terms of the medium in which each service operated and avoid duplicating weapons systems. Chief of Naval Operations (CNO), Chester W. Nimitz, believed that function, not weapon systems, should determine the composition and role of each military service. According to Nimitz, each service should be flexible and large enough to accomplish its mission. This view implied that the Navy and Marine Corps would retain their own aviation arms. Navy leaders strenuously resisted what they believed were Air Force attempts to exclude naval strategic air functions. The National Security Act of 1947 resolved this dispute in the Navy's favor, allowing the Navy to retain its own air force. However, the issues of roles and missions-the Navy's role in the strategic air offensive-remained a continuing source of conflict with the Air Force. [14] During 1946, for example, Navy leaders had tried (1) to dictate the bases, phasing, targets, weapons, and strength of the American air offensive and (2) to compete with the Air Force for the role of keeping the Soviets in check in the Mediterranean. [15] The 1947 act did not end that sort of competition.

The notion of the strategic air offensive was complicated further by the paucity of information available to policy makers about atomic weapons, the small number of weapons stockpiled between 1945 and 1947, and the weapons' physical size (see photographs of the Little Boy and Fat Man). During the early postwar period, there was very little communication within the US government about atomic weapons. Even policy makers with a "need to know" were not necessarily informed. From the end of the war through 1 January 1947, when the Atomic Energy Commission assumed control of all atomic weapons, research, and production facilities from the Army Corps of Engineers' Manhattan District, there were no formal requirements for the Manhattan District to report the size of the stockpile to top civilian or military leaders. Procedures for this transfer of information were first established in the Atomic Energy Act signed by President Harry Truman in August 1946. [16]

Fat Man

Between August 1945 and April 1947. President Truman was not briefed formally on the number of weapons in the stockpile. Furthermore. Secretary of State James Byrnes. Secretary of Navy James Forrestal, and the collective Joint Chiefs of Staff (JCS) leadership were not briefed either. The only formal line of communication came from Maj Gen Leslie Groves and Col Kenneth D. Nichols through General Eisenhower to Secretary of War Robert Patterson. There was no official procedure for either Eisenhower or Patterson to brief Truman. Truman eventually was briefed by AEC Chairman David Lilienthal on the state of nuclear production on 3 April 1947. [17]

The number of atomic weapons available was quite small. In June 1945, there were one gun-type uranium Mark I Little Boy and two plutonium-fueled Mark III Fat Man implosion weapons. The Little Boy was detonated over Hiroshima on 6 August 1945. One Mark III was tested at Trinity-the first test detonation of a nuclear weapon at Alamogordo, New Mexico, on 16 July 1945-and the other was dropped on Nagasaki on 9 August 1945 (see the Mark III test device. the "Gadget." photograph). On 30 June 1946, the stockpile contained between seven and nine Mark IIIs. Within the Air Force, Maj Gen Curtis E. LeMay assembled the first air war nuclear annex-detailing targets and tactics. He was unable to get information officially on the size of the arsenal until about August 1947. Secrecy concerning the number of weapons extended to the technical and physical characteristics- dimensions and weights-of the bombs themselves. Hence, important features of new aircraft designs, such as the size of bomb storage areas and of bomb bay doors, could not be determined with finality, and last-minute changes in one dimension would force designers to consider a new set of cascading engineering trade-offs. Thus, secrecy complicated the task of developing a nuclear capable air force, since many information restrictions were placed on personnel charged with the task of designing equipment and constructing nuclear capable aircraft. [18]

Navy officers devoted little attention to the use of the atomic bomb in 1946, primarily because of the many uncertainties about its capability and availability of atomic bombs. [19] Navy decision makers were concerned mostly about defending against an atomic attack on the fleet or on such targets as the Panama Canal. Further, submarine technology and antisubmarine warfare were urgent problems. Soviet capture of a number of advanced German U-boats (Type XXI and Type XXVI) posed a significant threat to the sea lanes. [20]

The first serious naval consideration of offensive use of atomic weapons appeared in the final report of the JCS Bikini Evaluation Committee in July 1947. The report stressed the need for an effective atomic striking force as a deterrent to attack. Nimitz extended this line of reasoning in his valedictory statement upon retiring as CNO, "The Future Employment of Naval Forces," delivered on 6 January 1948. Nimitz argued that the Navy had developed carrier technology and tactics to such a point that it could create offshore bases of superior capability and low vulnerability anywhere in the world. Further, the Navy was the service best prepared to project power against an enemy in the early phases of war, because a feasible intercontinental bombing force was not likely to be achieved for several years. Therefore, Nimitz argued, the Navy should be assigned the continuing responsibility to supplement Air Force bombing operations. [21] Nimitz's appraisal was not entirely correct. In early 1948 the on paper B-52 was still a turboprop design (Model 464-35), and there were reasonable doubts about its ability to achieve the desired range (range is the distance from takeoff to exhaustion of fuel supply). [22] However, the B-36A, which would enter the Air Force inventory in June 1948, could project power; depending upon payload, it had a range greater than 8,000 miles.

Other senior Navy officers did not agree completely with Nimitz, but they did support a greater Navy role in national defense. RAdm Daniel V. Gallery, assistant CNO for guided missiles, argued in a memorandum (dated 17 December 1947) that the Navy could handle most offensive air operations, including atomic attacks. RAdm Ralph Ofstie, a former member of the JCS Bikini Evaluation Committee and a member of the joint Military Liaison Committee to the AEC, wrote in January 1948 that the United States should emphasize development of high-performance, high-mobility Navy aircraft with improved accuracy, rather than the production of large numbers of land-based Air Force "super bombers." [23]

While there was little consensus within the Navy on naval doctrine in a future war, Navy leaders worked to keep nuclear options open. By 1946 naval aviation officers had seen that the combination of jet engines and atomic weapons opened the "possibility of adding true strategic strike capability to [the Navy's] other offensive roles, since it was no longer necessary- to think in terms of giant [land-based] superbombers for strategic operations." [24] The Bureau of Aeronautics outlined a proposal for a bomber able to operate from the large flush-deck carriers being planned. Douglas Aircraft Company received the contract for what became the A3D Skywarrior, and a suitable design for the aircraft was completed in 1949.

The Navy aircraft requirement was met by a swept-wing jet bomber, weighing 60,000 pounds (the largest and heaviest ever projected for use), and a large internal weapons bay with provision for 12,000 pounds of conventional or atomic bombs. [25] In 1949 an AMC briefing noted the development of this Navy bomber which could compete for the B-52's projected role. The Navy airplane's maximum speed of about 600 MPH, altitude of 41,000 feet, and radius of more than 1,000 miles exceeded the performance of the Air Force's XB-52 Model 464-35. (Since the Navy airplane would be carrier-based, its required radius could be shorter than that required for Air

Force aircraft based in North America.) [26] Air Force leaders and the public interpreted the Navy's efforts to develop the capability to wage atomic warfare as part of an attempt to gain organizational control of the air offensive. The conflict over roles and missions between the Air Force and Navy continued through 1947, and by the fall of 1948, was becoming even more intense than before the passage of the 1947 National Security Act. [27]

Thus, the B-52's development occurred in a period of great uncertainty and strife in the defense establishment where the attention of senior leaders was properly directed to broad issues. Peacetime had produced large cuts in personnel, budget, and orders to aircraft manufacturers. Although the destructive power of atomic weapons could help fulfill the "promise" of the strategic bombardment doctrine, little information was available about the bombs, only a few bombs were in the national arsenal, and these were large and heavy. Air Force leaders were preoccupied with a variety of critical issues related to independence from the Army and the establishment of a primary role in defense of the nation. They wanted a long-range bomber capable of carrying a 10,000- pound payload-not only to fight a war against the Soviet Union but also to help justify an independent role for the Air Force in the national defense. Against this background, many difficult questions surfaced during the first 13 months of the new heavy bomber program. But the senior people involved in acquisition of this new aircraft had little understanding of the size and complexity of their task. Most of their attention and time was devoted to broad political issues.

Douglas A3D Skywarrior

Ambiguous Technology and Changing Goals

No major Air Force organizational activity on the new bomber proposals took place until the middle of February 1946. A "design directive" calling only for "the greatest possible latitude in design" was distributed to the aircraft industry with invitations to bid on the military characteristics outlined in November 1945. Col George E. Price, chief of AMC's Aircraft Projects Section, viewed the ambitious required characteristics as a distant goal. Adequate power plants, he admitted, were lacking "at this time." [28] Boeing Aircraft Company, Glenn L. Martin Company, and Consolidated Vultee Aircraft Corporation submitted cost quotations and preliminary design data close to the requirements. [29]

Boeing responded to the request for proposal in April with its Model 462 (fig. I). This model weighed 400,000 pounds, had a tapered straight wing, and was powered by six. Wright Aeronautical Corporation XT-35 gas turbine engines combined with propellers. Correspondence between Boeing and AMC regarding the possibilities for future development of Model 462 was continually optimistic. [30] In a letter to Lt Gen Nathan F. Twining, commanding general of AMC, William M. Allen, Boeing president, boasted that the Model 462 "represents the optimum airplane possible in a balanced design based upon the design directive, and within the limitations of existing components and anticipated developmental considerations." [31] The mock-up airplane would be available for inspection on or before the end of August 1947.

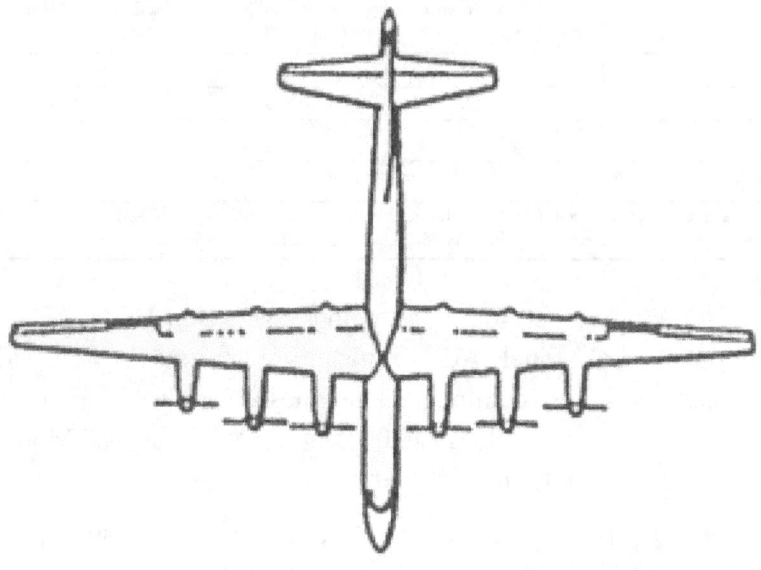

Figure 1. Boeing Model 462

On 23 May, in a letter to General Spaatz, Brig Gen Laurence C. Craigie, chief of AMC's Engineering Division, recommended awarding a single heavy bombardment aircraft contract to Boeing. [32] The recommendation was accepted and communicated to AMC on 29 May 1946, and three weeks later Model 462 officially became the XB-52. [33]

The Model 462 configuration spurred discussion among AMC personnel. A memorandum report prepared in June by personnel in AMC's Aircraft Projects Section predicted that Model 462 would experience many revisions in weight and balance, arrangement of crew, armament, navigation equipment, and similar items. [34] In August Col George E. Price wrote to Maj J. F. Wadsworth, Design Branch of AMC's Aircraft Laboratory, about the negative implications of aircraft weight. [35]

In September Maj Gen E. M. Powers (AC/AS-4, Materiel) echoed these concerns, arguing that a heavy intercontinental bomber could not be built without "prohibitive size unless other currently specific performance criteria are reduced." [36] Indeed, data regarding the relation between aircraft size and weight presented at an AMC conference in October posed a stark choice between military requirements and aircraft size. Airframe weight contributed only 29 percent of an airplane's maximum gross weight. The remaining 71 percent included power plants, armament, equipment, bombs, and fuel as directed by the military characteristics. Unless military characteristics were refined, a 5,000-mile radius bomber, flying at an acceptable cruising speed, and using T-35-1 turboprop power plants would weigh almost 500,000 pounds. [37] One reason air staff officers resisted the idea of such a heavy aircraft was that a 500,000-pound bomber would be too heavy to be supported by then-current airfields. Table 3 compares the maximum takeoff weights and combat radii of other bombers.

Table 3
Maximum Takeoff Weights and Combat Radii

Aircraft	Maximum Takeoff Weight (lbs.)	Combat Radius (miles)	Maximum Bomb Load (lbs.)
B-29	140,000	1.975	20,000
B-50A	168,480	2,190	28.000
B-36A	311,000	3,876	72.000*
*Later B-36 models could carry an 86,000-pound bomb load.			
Source: Marcelle Size Knaack, Encyclopedia of U.S. Air Force Aircraft and Missile Systems, vol. II, Post-World War II Bombers 1945-1973 (Washington, D.C.: GPO, 1988), 54-55, 200-201, 494.			

Within only six months of Model 462's acceptance, Maj Gen Earle E. Partridge's AC/AS-3 staff advised Boeing that the Model 462 six-turboprop engine design did not meet range requirements. [38] In response, Boeing commenced new design studies and produced a new design, the Model 464. Weighing approximately 170,000 pounds less than the Model 462, Model 464 was a 230,000-pound gross weight bomber powered by four turboprop engines. [39]

The major issues leading to the Model 464 proposal concerned range and s~. Both Powers and Partridge argued that a 400,000-pound gross weight airplane was too heavy. Partridge added that the range of the Model 462 would be too short to accomplish the desired mission. In response, Boeing proposed a much smaller aircraft, the Model 464, which some AMC officers felt would meet Air Force needs.

On 26 November AMC's Craigie recommended adopting the Model 464, but the next day another proposal carne out of a conference at Washington, D.C., attended by Generals LeMay, Powers, and Alfred R Maxwell and Boeing representatives. [40] The conference had been convened to decide the future direction of the XB-52 contract and to answer Partridge's criticism of the XB-52. Previously, Partridge had decided that Model 462 would not meet Air Force needs because it would be far too large and too heavy. Partridge wanted to defer XB-52 procurement for another year to gain more experience on the XB-35 (the Northrop Company's Flying Wing) and the B-36 Peacemaker and to put strategic planning on a firmer basis. [41]

As deputy chief of the Air Staff (DC/AS) for R&D, LeMay was deeply involved in Air Force planning for the use of atomic bombs and hence was concerned with developing aircraft needed to carry these weapons. He outlined the requirement "for a special task force of 5,000-mile [radius] airplanes capable of dealing a heavy blow from secure North American bases in the event that outlying bases were rendered untenable at the outbreak of war." [42] Partridge and Maj Gen Lauris Norstad, the assistant chief of Air Staff for plans (AC/AS-5), agreed to this proposal in principle. But LeMay also noted that some concessions on the requirements could be made:

• The plane should be designed specifically around the A-bomb without the usual full complement of conventional bombs, which in this case might have been 100,000 pounds for short range operations. The disposable bomb load could still reach 20,000-30,000 pounds without compromising the design.

• The aircraft's mission should be "special operations and not sustained operations, and therefore the advantage of surprise" would permit the installation of tail armament only instead of all-around armament. There would also be no need for a parasite fighter.

• The new crew size should be reduced to a minimum to achieve success in the "special mission."

• The size of the force should be about one wing, minimizing the need for new airdrome construction.

• The aircraft should have a minimum operating radius of 5,000 miles and a reserve of 2,000 miles to accomplish the special mission from North American bases. The total range, with 10,000 pounds of bombs-the weight of a Fat Man Mark III atomic bomb-would be 12,000 miles. This range would be an improvement of 2,000 miles over the B-36. The possibility should be considered of dropping a portion of the landing gear after takeoff, leaving only sufficient landing gear for the pilot to land practically empty.

These requirement concessions completely altered the Model 462 conception of the B-52. LeMay had moved away from the general purpose, flexible, completely armed, alternate range bomber. LeMay's vision of the B-52 also was different from the Model 464 approved a day earlier at Wright Field.

Boeing representatives argued that a number of these suggested changes required staff study by the Air Force, design study by Boeing, and a slower incremental approach to design. They suggested that new design problems, discovered in 1946, made it desirable first to build a high-speed, medium bomber as a successor to the B-50 before moving to a 5,000-mile radius bomber. Believing they could rally political support for a heavy bomber more easily than for a medium bomber, Air Force officers rejected this suggestion. Eventually, all conferees agreed to change the XB-52 contract to a design study along the lines of LeMay's requirements. This decision postponed the necessity of asking Chief of Staff Spaatz whether funds should go to a special mission bomber or to a prototype successor of the B-50. They recognized that the knowledge necessary to choose was not available. On 7 December Powers informed Twining of the decision reached at the 27 November conference held in Washington, D.C. Powers added that Boeing should initiate a study for the design of a bomber to meet requirements for accomplishing a long-range atomic attack. [43]

Boeing B-50 Superfortress

While officers on the Air Staff were changing their vision of the next long-range bomber, discussions proceeded at AMC on the turboprop propulsion system. The Wright XT-35 gas turbine had been proposed for the XB-52 in August 1946. The engine's development was as uncertain as the airframe design, with many disagreements about the reduction gear and propeller. Representatives from Wright and Boeing disagreed about XT-35 reduction gear design features. The need to develop a reduction gear was a result of simple engineering realities: the different efficiencies of turbines and propellers. Propellers operated efficiently at speeds considerably lower than turbines could generate. Hence, the reduction gear was necessary to combine the power generated by the turbine with the propeller. Personnel in AMC's Power Plant Laboratory estimated the new reduction gear would cost about one million dollars and would take one year to complete. [44]

AMC staff stressed, too, the importance of "proper planning" of propeller development so that tested and proven propellers would be available ahead of the delivery requirements of the XB-52. This goal implied the successful coordination and completion of a difficult and time-consuming set of tasks: propeller research and development involved three to four months for a preliminary aerodynamic study and then six to eight months to develop the detailed design, assuming the basic principle of the propeller actuating mechanism (governing hub pitch and speed and propeller pitch) was proven. This stage would take longer if new actuating features had to be incorporated. An additional seven months were required to deliver a testable propeller, followed by six months to deliver a quantity of propellers sufficient to supply one airplane. Thus, the amount of time devoted to propeller development, optimistically assuming no delays, would be at least 22 months.

In 1946 not even the basic feature of the XB-52 propeller- its diameter-had been fixed. The Curtiss Propeller and Hamilton-Standard Propeller companies promised to continue studies and return to Wright Field in October 1946 for more discussions. Connecting the propeller to the engine added difficulties. Propeller manufacturers claimed to need two engines each for testing. Yet, AMC's Power Plant Laboratory expected the first T-35 gas turbine would not be available for flight testing until late 1947. Also the number of engines on order from Wright were insufficient to furnish all claimants: the propeller manufacturers, Propeller Laboratory (AMC), and Power Plant Laboratory (AMC). [45]

For several years, power plant engineers had recognized demanding developmental problems posed by turboprop propulsion systems. In 1939, for example, a GE study concluded that a turbojet would be preferable to a turboprop as a way to use the gas turbine. As noted in the last chapter, the technical obstacles to a turboprop were so great that none was in service as late as mid-1949-despite the great resources devoted to this type of engine during World War II. [46] In 1946 the technical problems entailed by the proposed turboprop propulsion system had begun to manifest themselves to AMC managers, and these obstacles created new difficulties for Boeing as the airframe designer. The first design studies for the Wright T-35, a centrifugal-flow gas turbine, were completed at the beginning of 1945, and the first use of the engine projected in the B-36. This use was abandoned, and since it was projected to be available at a critical period during aircraft development, the T-35 was adopted for the B-52 as Boeing completed design studies in early 1946.

Each party concerned with propulsion-Wright, Boeing, and the propeller manufacturers-introduced competing demands or obstacles, all with the potential to delay delivery schedules or impair aircraft performance. The T-35 design was revised several times, predictions of increased power and lowered specific fuel consumption accompanied each change. [47] By January 1947 no component construction had begun for the T-35 engine. Total work consisted of general design study and some detailed component drawings. In 1947 the T-35 (and a back-up program, T-37) absorbed approximately 60 percent of the year's Air Force engine development budget. [48] Simultaneous development efforts were made to construct testable units of the T-35 compressor combustion section and turbine, and its propeller, drive shaft, gearbox, and power control components. The difficulties in constructing the latter four components were greater than had been predicted. Yet, considerable optimism existed that a turboprop system could be developed. George Schairer, Boeing staff engineer, reported that Wright predicted the revised T-35-3 would increase available power at altitude by over 50 percent. With increased power and lower specific fuel consumption, a specially designed propeller to achieve increased range for the B-52 would be less important. [49]

The Air Staff and AMC: The Impact of Different Decision Premises

During the first 13 months of the XB-52's program, the technologies-engine-propeller and airframe designs-were not yet available to achieve the stated Air Force requirements. By the beginning of 1947, some information about Boeing's designs was clear (e.g., projected high airframe weights) and raised major concerns-the projected weight of designs submitted by Boeing conflicted with existing Air Force facilities. Other information was imperfect or not yet available, for example, whether the turbine-propeller combination would meet range and speed performance projections.

Air Staff and AMC officers approached B-52 development differently in response to the quality of information available to them and the amount of time they had to work on the program. On the one hand, senior Air Staff officers tended to revisit and revise requirements. But, their attention was distracted by other issues, including independence, mobilization, training, roles and missions, acquisition of other aircraft (such as the B-36), and competition with the Navy. Air Staff officers viewed the B-52 as a potential solution to such concerns as competition with the Navy for the mission of strategic attack and advocated range, speed, and payload requirements that supported the B-52 in that role. Thus, much of the information and requirements generated by Air Staff officers was framed more by the political problem 0: convincing members of Congress that the Air Force could execute the strategic mission than by the desire to clarify the technological options. On the other hand, AMC officers, less attuned to politics and assuming incremental design improvements would be made once the aircraft was in service, were committed to Boeing's designs. They resisted reappraising the match between requirements and aircraft design, which could reopen the design competition. Such an action could throw the B-52 program in jeopardy, since a potential outcome was termination and development of an aircraft designed by another firm.

The interaction of these differing perspectives had the unanticipated effect of enhancing critical scrutiny of requirements and technology. It spurred AMC officers to continue to investigate whether new technologies and designs could meet the objections and requirements posed by their counterparts on the Air Staff. In effect, the interaction of the Air Staff and AMC, aided by the private airframe, engine, and propeller firms, created a small multiorganizational system capable of culling the best option from a large number of candidates.

The story of AMC and Air Staff interaction over the B-52 also was influenced by frequent rotation of senior Air Force officers, reducing the amount of time any officer could devote to becoming expert in particular issues and making for a constantly changing cast of characters. The appendix provides a select list of positions and length of tour for critical officers involved in the B-52 program, and shows how personnel who influenced B-52 development moved into and out of the program.

At the next B-52 conference held at the Pentagon in January 1947, representatives from Boeing and six different Air Staff offices, including LeMay and Powers, attended. Boeing presented two general aircraft configurations to meet the requirements LeMay had outlined in November 1946. The first design, Model 464-16, was a tapered straight-wing airplane powered by four turboprop engines (T-35 gas turbines). The designers predicted it could carry 10,000 pounds of bombs at least 12,000 miles while cruising at 420 MPH. Without armor or defensive fire, the Model 464-16 would have a 13,800-mile range. No provision was made in this version for conventional bomb loads; the remainder of the fuselage would contain permanent fuel tanks. Model 464-17 was similar to 464-16, but it could carry a 90,000 pound conventional bomb load. Both planes had a predicted gross weight of 400,000 pounds-the same weight as Model 462. [50]

Most conference participants seemed pleased with the, aircraft's predicted performance. Eventually, Model 464-17 was selected and funded under the original XB-52 contract. Nevertheless, two subsequent technical complications would have political implications for the Air Force position on roles and missions. First, progress was slower than expected on the Wright T-35-3 gas turbine chosen for Models 464-16 and -17. The less efficient T-35-1 would have to be used until the T-35-3 became ready, increasing uncertainty regarding the achievement of range. The predicted range for the T-35-3 powered B-52 had to be based on an extrapolation from the T-35-1 ratio of propeller efficiency to specific fuel consumption. In the event the T-35-3 proved a failure, the actual range achieved with the T-35-1 would fall short of the Air Force requirement.

Second, the Model 464-17 (fig. 2) design had no provision for fuel protection. It was clear that the strategic mission would be vulnerable to any enemy action leading to punctured fuel tanks. [51] The question of fuel protection posed a difficult trade off. Obviously, flying with fuel tanks unable to sustain damage would reduce the probability of mission success. Yet, the added weight of leak-proof fuel tanks decreased range. Failure to achieve desired range, as specified in negotiations among senior officers on the Air Staff, would entail serious political risks for those justifying the B-52 to critics inside and outside the Air Force. Consequently, Brig Gen Alden R. Crawford, chief of the Research and Engineering Division in the Office of AC/AS for Materiel, asked Twining to prepare fuel tank safety requirements leading to the least possible reduction in range.

Discussions of the requirements for heavy bombardment aircraft continued in Washington as various studies conducted at AMC and Boeing showed the complexity of the project, addressing critical topics ranging from landing gears to runway thickness. [52] Research at Boeing, Wright, and AMC laboratories throughout the first half of 1947 proceeded with little reference to Air Staff disputes over military requirements for the B-52. Some of this research focused on gas turbine engines and their integration with the fuselage. Boeing delivery schedules were based on receipt of working engines by certain dates. William M. Allen, president of Boeing, predicted confidently that the first flight of the B-52 would occur in January 1950. [53]

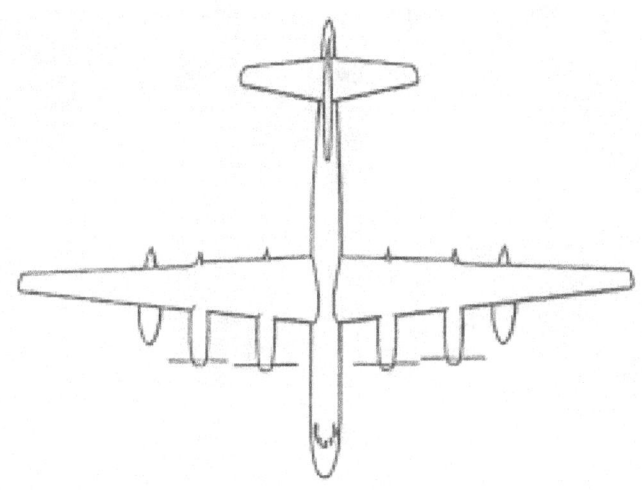

Figure 2. Boeing Model 464-17

Air Materiel Command staff emphasized the high priority of the B-52 and sought to ensure uninterrupted funding to continue research. [54] But Air Staff personnel were not sanguine about the B-52. Representing the office of AC/AS for Materiel, Crawford replied to AMC that initial research had not proceeded sufficiently well to justify its continuation. [55] Uncertainty at AMC over funding was resolved by the Air Staff, but permission to initiate further research depended upon Maj Gen E. M. Powers's approval. [56]

In Washington, Partridge's staff tried to expand the range of aircraft options appropriate for the long-range mission. They continued to search for ways to increase the B-52's speed and range, and to reduce its weight and size, but they also examined the proposals of other airframe firms. [57] By March 1947 Boeing had proposed several aerodynamic improvements to Model 464-17, changes which led to Model 464-29 (fig. 3). The new model had a predicted cruising speed of 455 MPH-almost 30 MPH faster than Model 474-17. However, at the end of April, Brigadier General Maxwell (chief, Requirements Division, AC/AS for Operations and Training) suggested that Model 464-29 suffered from three technical difficulties relating to range. First, the Boeing prediction of range, based on extrapolations of the T-35-1 ratio of propeller efficiency to specific fuel consumption, was too optimistic. Second, he cited a Project RAND-Douglas study which suggested that Boeing's range prediction was wrong. Finally, he noted Boeing's determination to use a conventional straight-wing fuselage design. The latter decision remained firm despite evidence that two new designs-the delta wing or Northrop's XB-35 Flying Wing-promised greater aerodynamic efficiency. The available performance data, from paper studies and wind tunnel tests, argued for a smaller B-52-without bomb bays and with weapons carried in external pods. [58]

Brigadier General Crawford was considerably more cautious in using Northrop and RAND studies to question the B-52 program. He argued that the RAND data showed "trends rather than detailed finished information." The Research and Engineering Division tried to get Boeing, Douglas, and RAND personnel to explain and resolve differences in their respective studies and projections and to solicit comments from AMC. [59]

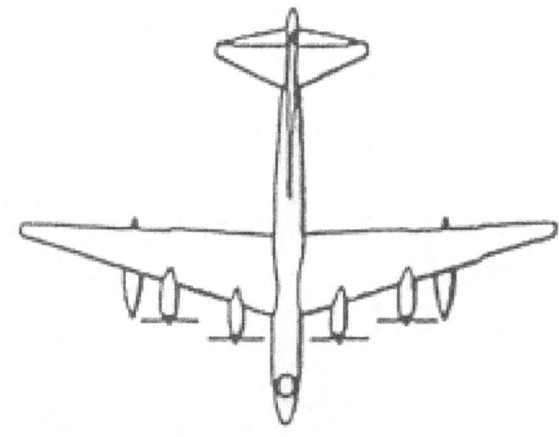

Figure 3. Boeing Model 464-29

In addition to these exchanges regarding the B-52 requirements at the Air Staff level, similar discussions took place between high-level Air Staff and AMC officers. AMC's Twining was deluged with letters from senior Air Staff officers about requirements, the state of technology, and the best way to contract for a heavy bomber. Powers wrote to Twining about the disputes over the RAND studies of B-52 range. Less cautious than his subordinate Crawford, Powers argued the RAND weight projection curves showed that if the B-52 were overweight by even 2 percent, it would miss its performance goals in cruising speed, range, and payload by a wide margin. The RAND studies he wrote "indicate that a Delta Type design is far more satisfactory for a long-range, high speed airplane than the conventional fuselage presently visualized." [60] The Air Force could ill afford to ignore the doubts raised by the RAND reports, Powers concluded. He urged AMC to study the issue of B-52 range performance and provide the Air Staff with the results immediately. [61]

LeMay, in his position of DC/AS for R&D, prepared the first air war nuclear annex detailing targets and tactics for the Air Force. With knowledge of the plans for the conduct of nuclear operations, he advocated those requirements for new bombers that would contribute to the Air Force's ability to wage atomic warfare. Hence, LeMay restated concerns regarding range and requirements to Twining, noting the many discussions held in Washington in the attempt to develop a program for the next five to 10 years "worthy of being presented to General Spaatz for consideration." The lack of an experimental program during World War II, LeMay predicted, meant that "new and improved bombers" would not be available for seven to 10 years. Further, he prophesied that the Air Force could not procure heavy bombers in great enough quantity to replace the B-36s and B-50s until the 1952-54 period. The B-52 was likely to be expensive, allowing only about 100 aircraft to be purchased during the life of the program. [62]

LeMay reiterated the broad goal of the heavy bombardment program with respect to the Air Force role in executing atomic strikes from US territory. The B-52 would carry an atomic bomb and "would be used to initiate retaliatory combat operations from North American bases, as soon as hostilities start. Such operations would obviously be on a small scale, due principally to the limitations of the atomic bomb stockpile, limited number of carriers available and the complications of operating at such long ranges from the U.S." [63] LeMay maintained that the B-52 should have a radius of at least 5,000 miles, carry a bomb load of 10,000 pounds, and be able to cruise at a relatively high speed. On paper, Boeing's Model 464-29 appeared to be that plane, but this choice depended upon the successful development of the T-35-3 engines.

In early May 1947, General Powers argued that if the B-52 failed to meet Air Force strategic needs, it raised the question of whether the heavy bombardment program should be oriented to the future or to modernizing existing aircraft. [64] LeMay urged the Air Force to look at bomber proposals from Consolidated Vultee, Northrop, and Douglas to ensure that Boeing's B-52 was the plane "we want to back," but he suggested the Boeing contract be continued for at least six months to allow consideration of other proposals. [65]

In June 1947 Air Staff officers again urged Twining to take stock of the entire B-52 program. A detailed memorandum prepared in AMC's Aircraft Project Section reviewed the history of the B-52 project, and outlined four issues related to the central problem of the B-52's range. [66] First, Air Staff officers had not understood the relation between size and range and how only a large airplane could achieve the requirement for a 5,000-mile radius. Second, studies by AMC, Boeing, and RAND showed that the range of the T-35-3 powered B-52 was on a curve of rapidly diminishing marginal returns. Even if Air Force officers could agree on aircraft size, it would be difficult to meet a 5,000-mile radius requirement with the turboprop power plant. Third, the basic

B-52 design was well-balanced, combining long-range flying with high average cruising speed. Range requirements could be met with weight reductions. Finally, Air Force officers had to assume that during development, engineering compromises would arise, resulting in a temporary range reduction. With proper planning, later models could achieve the required range.

The AMC memorandum recommended improving the B-52 development program by reducing the technical and political uncertainties generated by disagreements between the Air Staff and AMC:

• Evaluate options regarding weight of various items and secure agreement among AMC, AC/AS 3, and AC/AS 4 on those items for the B-52. Institute a weight control program with Boeing.

• Select the minimum weight "special mission" configuration (Model 464-17).

• Have the Air Staff outline, in writing, the basic design goals of the B-52, rank the B-52 against other Air Force projects, and exempt the B-52 from "automatic compliance with blanket directives-written for aircraft."

• Allocate the funds to assure development of the T-35-3 turboprop to obtain the desired specific fuel consumption. Provide a back-up power plant program for the T-35-3.

• Initiate a R&D program by NACA, AMC, and industry to assure a thorough investigation of propellers operating under conditions expected for the B-52.

• Do not penalize the design of the B-52 by making provisions for either an 80,000-pound bomb load or the internally stowed fighter.

• Have a prototype B-52 available for flight in 1950. Select an aircraft configuration and do not interrupt gas turbine and propeller development programs. [67]

This memorandum report was released from AMC the same day the Air Staffs Maxwell issued new "minimum acceptable" military characteristics (listed in table 4) for the B-52 in Washington. D.C. [68] After these basic requirements were met, Maxwell maintained, attention should be directed to high speed, armament, and passive protection to reduce vulnerability in penetrating heavily defended zones.

The predicted performance of Model 464-29-not the Model 464-17-approximated the requirements outlined in table 4. Yet, this aircraft would weigh about 400,000 pounds and the concern over high weight had not abated. The new military requirements differed from those issued in November 1945 in both speed and mission. The minimum acceptable speed had risen from 300 MPH to 420 MPH, and the mission had become limited to an attack with a 10,000-pound atomic bomb.

Table 4
XB-52 Performance Requirements

Performance	Minimum Acceptable
High speed at tactical operating altitude for 15 minutes	420 MPH
Tactical operating altitude	35,000 feet
Service ceiling	40,000 feet
Service ceiling ½-engines	15,000 feet
Tactical operating radius with 10,000-pound bomb load and full fuel	5,000 statute miles
Average speed for above radius	420 MPH
Takeoff over 50-foot obstacle at design gross weight without jet-assisted takeoff (JATO)	7,500 feet
Landing over 50-foot obstacle at design gross weight less droppable fuel and bombs	7,500 feet

A major threat to continuing B-52 development was uncertainty about whether it would meet its range requirements. In May 1947 LeMay had argued (according to Craigie) that if the B-52 fell below a 5,000-mile radius for any reason, the Air Staff should be notified, implying that the Air Force would cancel B-52 development. The range requirements of the B-52, Craigie admitted to Spaatz, probably would fall below the 5,000-mile radius, but he expected this deficiency could be resolved during the life cycle of the airplane; experience with both the B-29 and B-36 supported that view. He added that both bombers would have been canceled had the suitability of either been based on a paper design rather than a completed experimental airplane. Craigie maintained the deficiencies of the B-52 were temporary and should not obscure the potential of the design. [69]

AMC officers responded to criticism of the B-52's range from the Air Staff and RAND, arguing that oversimplified assumptions prevented accurate projections of airplane performance, Craigie cited a 16 May 1947 letter from Arthur E. Raymond, Douglas Aircraft Company chief engineer and early supervisor of Project RAND, who admitted that,

> RAND did not consider the effect of different cruising speeds. ...After allowance was made for these differences the so-called discrepancy disappeared. ...Boeing's figures for the B-52, so far as we know them, while imposing an admittedly difficult design problem, are feasible of accomplishment, provided the engines achieve the power and fuel consumption promised by the manufacturer, [70]

Craigie added that AMC analyses of other studies of optimum airplane performance relied on extrapolations from past performance. However, these extrapolations were unreliable because aeronautic and aerodynamic knowledge was growing so quickly. Therefore, it was necessary to use the most current data and knowledge, which did not necessarily involve only extrapolations of past performance. For instance, he noted, that in 1941 Douglas Aircraft Company analyzed the predicted performance of the B-36, and concluded that the requirement of a 10,000-mile range with a 10,000-pound payload was unlikely to be achieved. The Air Force and Convair, however, used improved weight control and planning, and proved the study wrong. Despite increases in armament, radar, equipment, and the difficulty of development under wartime conditions, Convair and the Air Force produced an airplane which could meet the Air Force's objectives. [71]

Craigie also urged discarding the alternatives to the B-52-the XB-35 and YB-49 Flying Wing and delta-wing designs. By May 1947 the delta wing did not have any marked superiority over a conventional airplane for long-range, high -speed operation. Craigie wrote that a reevaluation of these designs should be made only when "jet engine specific fuel consumption is reduced to a point to permit their [sic] use" in a bomber. [72]

Northrop's YB-49

These arguments failed to protect the B-52, because the Air Staffs concern was political; an aircraft design was needed to defend the Air Force's role in national defense. The Air Staff feared that the B-52's performance shortfalls would be used by the Air Force's bureaucratic opponents to enhance their own national security roles, In a letter to Twining and Craigie, LeMay suggested imminent changes in the B-52, noting that the B-52 program was only a study of "one method of accomplishing the strategic mission intended for this airplane. This project must be carefully and continuously scrutinized to assure its continuing practicability." [73] LeMay added that any directive to Boeing to begin actual construction would depend on the express authority from JCS and the Air Force commanding general. Indeed, LeMay suggested that AMC should begin to study other ways to accomplish the bombardment mission assigned to the B-52, including one-way flight to the target, the aircraft's return to friendly territory outside the United States, planned ditching areas, and use of subsonic pilotless aircraft. He added, "the strategic mission remains firm but the method of its accomplishment is not fixed. Economic considerations may force adoption of some other method-of satisfying the long-range requirement." [74]

Over the next few months, there was little official correspondence regarding the B-52 between AMC and Air Staff. The Air Staff was busy with service unification and administrative reorganization of the Air Force. Research activity on aircraft design continued at Boeing in conjunction with the subcontractors and AMC. Boeing proposed changes in propeller design, design of landing gears, amount of defensive armament, fuel cell construction, and size of bomb load. [75] Air Force officers at AMC tried to keep abreast of the vast amount of research activity. Their task was complicated by a casual attitude toward the exchange of design data among firms working on the B-52: information concerning discussions and decisions among manufacturers was not always passed on to officers at AMC. [76]

Shortly after the 1947 National Security Act became effective, the Committee on Long Range Bombardment, composed of representatives from the Air Staff, SAC, AMC, and Air University, met at the instigation of the Aircraft and Weapons Board to review heavy bomber military characteristics and make recommendations. The Aircraft and Weapons Board, composed of the Air Force deputy chiefs of staff and the major air commanders, reported to the chief of staff and met as major issues surfaced that required high- level consideration. The Long Range Bombardment committee recommended amending heavy bomber requirements and asked AMC's Aircraft Laboratory to prepare a study of an airplane to meet new military characteristics, including 8,000 miles range, 550 MPH speed, (550+ MPH over defended area), droppable landing gear, full purging and self-sealing tanks for fuel, crew of five (pilot, relief pilot, navigator bombardier, weaponeer, gunner), and air refuelability. The committee members recognized that the Aircraft and Weapons Board would not be likely to accept the need for self-sealing fuel tanks, droppable landing gear, or cruising speed as high as 550 MPH, so they were prepared to be flexible on these characteristics. [77]

The Committee on Long Range Bombardment's recommendation represented a substantial departure from the Model 464-29 version of the B-52, especially the reduced crew size and increased cruising speed. The speed requirement had been increasing steadily: from 300 MPH in November 1945, to 420 MPH in June 1947, to 550 MPH in October 1947. The 464-29 also would have had a range of at least 12,000 miles to achieve a 5,000-mile operating radius. The new characteristics reduced range to only 8,000 miles-hence the unrefueled radius would be reduced too. Agreement among all parties on these changes in military characteristics was easier to achieve because of LeMay's absence; as the newly appointed commanding general, United States Air Forces Europe, he was not, for the moment, involved in B-52 development.

By mandating a different configuration of the B-52, the committee's proposed changes initiated a new set of technological trade-offs. Officers in the newly organized Office of the Deputy Chief of Staff, Materiel (DCS/M) tried to secure agreement with Lt Gen Lauris Norstad, the deputy chief of staff, operations (DCS/O), matching a new B-52 configuration to the revised military requirements. [78] Maj Gen L. C. Craigie, formerly chief of AMC's Engineering Division and now in the office of DCS/M (headed by Lt Gen Howard A. Craig), noted,

> The Committee on Long Range Bombardment has concluded that the XB-52 in its present configuration does not present a practical solution to the long range bombing problem, and has recommended that the XB-52 be changed from its present configuration to an airplane having 8;000 miles range, and other characteristics.

Craigie suggested that AMC first should be directed to withhold further expenditures on the Model 464-29 (XB-52). Then, if the Aircraft and Weapons Board approved the new military characteristics, the B-52 contract would be change ordered. The new airplane would retain the B-52 designation for convenience in accounting and budgetary considerations. [79]

Two weeks later, the Aircraft and Weapons Board's Bombardment Subcommittee met to consider the military requirements for strategic bombers. Craigie attended from DCS/M and Frederic Smith and Partridge represented DCS/O. Smith opened the meeting with an Intelligence

Division report stating that the Air Force "could not expect to operate from bases 2,000 miles from the target and that it would be necessary to have airplanes of at least 3,000 miles radius." A 3,000-mile radius translated into an 8,000-mile range. The airplane would be over enemy territory-the Soviet Union-for a great distance, hence the mission cruising speed should be 550 MPH. [80]

The problem with the new cruising speed requirement was in the trade-off with range. RAND and AMC studies suggested that an aircraft capable of flying a 3,000-mile radius/ 8,000-mile range could be designed, but only at a cruising speed of 500 MPH. A great deal of time and money would have to be devoted to develop airframes, engines, armament, and propellers before any of the performances demanded by the new military characteristics could be met. Partridge noted that the matter of bases for operation and targets had been studied and discussed thoroughly at the last Aircraft and Weapons Board meeting in August 1947, and a decision made to retain the military characteristics for the "workhorse bomber" rather than adopt the 3,000-mile radius, 550 MPH airplane. Partridge argued that the "workhorse bomber" was the best aircraft for development. Smith agreed with Partridge and stated he would recommend to Craigie at the Directorate of R&D the cancellation of the Model 464-29 and the development of an airplane capable of 500 MPH cruising speed, 8,000-mile range, and weighing approximately 300,000 pounds. The meeting adjourned with the decision to request a study of heavy bomber parameters by AMC laboratories. [81]

The next day, 19 November, Partridge approved Craigie's plan to stop development of the Model 464-29 version of the B-52. He suggested using the new military characteristics for an 8,000-mile range heavy bomber approved in October instead of the B-52 characteristics of June 1947. The June 1947 characteristics had called for a bomber capable of carrying a 10,000-pound bomb internally, while flying a 5,000-mile radius, at 420 MPH, and weighing about 400,000 pounds. The new military characteristics seemed to exploit best the potential of the all-wing type aircraft, that is the Northrop Flying Wing. Therefore, he recommended opening the contract for the airplane to meet the new military characteristics of 8,000-mile range and 500 MPH cruising speed to all interested firms instead of giving it directly to Boeing. However, the new military characteristics had not yet been seen by the Aircraft and Weapons Board members. The characteristics would have to be circulated to them, and the next meeting would take place on 20 January 1948. [82]

While Aircraft and Weapons Board members were considering new heavy bomber characteristics, lower level Air Staff officers continued to discuss military requirements (see the appendix to review the experience and position of each officer). In a letter to Craigie, Col Clarence S. Irvine, assistant to the chief of staff of SAC, noted that since preparation of the "Report on Heavy Bombardment" in November, he had continued to study the heavy bomber concept in connection with work at SAC on two B-50 mock-up boards. This work led to new data and a possible reappraisal of heavy bomber capabilities according to four criteria. First, he suggested that the new military characteristics for heavy bombardment aircraft be written around a minimum speed of 550 MPH. Second, he reported that although information available to the Aircraft and Weapons Board indicated poor propeller performance at speeds over 450 MPH, more recent studies by propeller companies showed that blade efficiencies over 80 percent could be attained at speeds up to 540 MPH. Third, he noted that an 8,000-mile range would allow the bombers to attack a target up to 5,000 miles away and return to an American base with only one refueling. Studies of jet aircraft suggested that fuel consumption would limit range to about 6,000 miles for aircraft weighing less than 300,000 pounds. In terms of fuel consumption, the state of jet engine development could not lead to an airplane of reasonable size capable of 550 MPH speed and 8,000-mile range. [83]

Fourth, Irvine suggested an aircraft design that was discussed informally with Boeing engineer George Schairer. The airplane would have a range of 8,000 miles, a speed of 520 MPH over 4,000 miles of enemy territory, a wing area of 2,400 square feet, tail armament only, and a gross weight- with a 10,000-pound bomb load-of 280,000 pounds. Several aspects of this design would be taken from the advanced model of the B-50C. Hence, this design would require relatively little research to achieve. Irvine concluded his letter to Craigie with an appeal for greater attention to turbojet engines in the Air Force research program and attached three charts, each detailing a different airplane, all with the same turbojet power plant. [84]

Craigie also received a letter from Col J. S. Holtoner, chief of Aircraft Branch in the Office of DCS/M, dated the same day as Irvine's. Holtoner's letter suggested major changes in the B-52 program. First, Holtoner recommended canceling Model 464-29 because it would be too large and too heavy, could not attain planned range with its designated power plants, and would be obsolete before its completion. [85] His proposal for a heavy bomber, Holtoner admitted, would reduce operating speed from 550 MPH to 500 MPH. However, it would embody very recent aeronautical advances and would represent an overall improvement in performance over the Model 464-29, and would mesh well with the Boeing B-52 program. Holtoner concluded that a better airplane would result if the B-52 contract were opened for competition. If Boeing should win the competition there would be no loss of funds. If Boeing should lose the competition, the financial losses would be offset by more advantageous cost figures from the new contractor. Holtoner closed with the recommendation that new characteristics for the B-52 be circulated to the aircraft industry for competitive proposals and that these decisions be submitted to the Aircraft and Weapons Board for approval. [86]

Discussions concerning the B-52 requirements continued at the Air Staff. On I December, a meeting attended by Generals Craigie, Chidlaw, Powers, Carroll, Crawford, Smith, and Partridge reviewed a report on the B-52 prepared by a civilian consultant, Courtland D. Perkins. Based on Perkins' findings that it would be too slow and too expensive, they agreed unanimously to stop further expenditures on the Model 464-29. The Perkins Committee report recommended that the military characteristics of a replacement plane should include cruising speed of 500 MPH, range of 8,000 miles, and gross weight of less than 300,000 pounds. Attainment of the range requirement would not be guaranteed. The range figure exceeded that given in AMC studies but agreed with figures given by RAND and Boeing studies. [87]

A meeting was held the following day at Craigie's office to discuss several different aspects of the B-52 program. It was agreed that the new military characteristics should include an 8,000-mile range at 500 MPH cruising speed. Craigie noted that Boeing, RAND, and Northrop studies indicated that a bomber capable of such characteristics would weigh approximately 300,000 pounds. AMC representatives disputed those predictions and stated that the range of the replacement plane would be only about 7,500 miles. At the same meeting, AMC's Lt Col Heruy E. "Pete" Warden compared gas turbine and turbojet engines for speeds of about 500 MPH and concluded that without further propeller development turbojet engines were more feasible. [88]

New range, speed, and weight requirements doomed Model 464-29, and as a result of the discussions on the Air Staff, Vice Chief of Staff Hoyt S. Vandenberg wrote to Secretary of the Air Force W. Stuart Symington proposing to cancel the Boeing XB-52 contract. [89] Symington directed terminating the Boeing contract and opening a new competition for a heavy bomber. [90] Informal word regarding the cancellation was sent to Boeing, and Boeing President William M. Allen protested to Symington. Allen argued that a new competition would cost the government extra time if Boeing won the competition, and extra time and money if it did not. [91] In either case, the options were poor from the government's standpoint. McNarney also requested reconsideration of the decision to cancel the B-52 contract and open a new competition. [92]

The cancellation of Model 464-29 posed stark challenges for AMC and its analytical style. The use and production of aeronautic and aerodynamic knowledge were integral to the way AMC conducted its business. The pace of activity for the AMC managers of B-52 R&D had been intense throughout 1947. Air Materiel Command officers and laboratories dealt with fairly specific questions in the attempt to make better informed procurement decisions about aircraft. For instance, would an appreciable increase in range result from the use of a droppable landing gear? Or, how would parasite fighters affect the achievement of range? Or, is turbojet propulsion feasible for heavy bombardment aircraft? AMC officers, more than Air Staff officers, acknowledged the limits of their knowledge of aerodynamics and aeronautics, and they sought experimental evidence to evaluate aircraft designs. During 1947 AMC officers organized a "fly-off competition" of turbojet powered medium bombers. At that time, little was known about the potential low drag and high performance of swept-wing aircraft. [93] It was discovered that the XB-47 had only 75 percent of the drag that had been optimistically predicted for it. The theories underlying the paper studies of the XB-47 had completely underestimated the potential for drag reduction found when the experimental swept-wing airplane was actually flown. [94]

AMC laboratory officers recognized such technical uncertainties in developing a heavy bomber. Nevertheless, in early 1947, they were optimistic about the ability of conventional aircraft designs and technologies. At the same time, they made a conscious attempt to base design and procurement recommendations on empirical evidence, rather than on wishful thinking. For example, despite having been involved with jet propulsion for several years, Craigie, while assigned to AMC, cautioned that turbojet engines should be reconsidered only when specific fuel consumption allowed achievement of longer ranges. In 1949, two years after Craigie's caution, the J57 turbojet engine (which made commercial jet transport economically feasible) demonstrated a much better specific fuel consumption.

Throughout 1947 conflict erupted between AMC's B-52 managers and Air Staff officers evaluating heavy bomber design and requirements. This conflict was rooted in the nature of the Air Staffs discussions about requirements in which they attempted to reach consensus about abstract military aircraft characteristics and the employment of aircraft in war rather than to solve hardware problems. Under the surface, senior Air Staff officers faced symbolic and political problems; they could not approve a heavy bomber (and associated military characteristics) that would prove a liability in public debates over the Air Force's role in national security. As noted above, in late 1946 Air Staff leaders rejected Boeing's recommendation to substitute a high-speed medium bomber for the B-52, because they believed a heavy bomber was more likely to win congressional approval.

Hard bargaining, as between the AMC and Air Staff, placed strains on an organization's status and power relationships. By legitimizing different organizational goals, it compromised the primacy of coordination by the organizational hierarchy. Hence, when bargaining occurs in any organization, it is frequently concealed in an analytic framework. [95] AMC officers could not enforce their views of what military characteristics to approve or what aircraft to purchase. Air Staff officers lacked the training and education necessary to understand the nature of technical questions they asked about developing high-performance aircraft. For example, LeMay told Twining that new and improved bombers would not be possible for seven to 10 years because of the state of aerodynamic research. Although an acclaimed combat leader, LeMay was not qualified to evaluate the state of aerodynamic knowledge. (A new and improved medium bomber, the XB-4 7, flew approximately eight months after LeMay offered his gloomy prediction to Twining.) Thus, Air Staff officers requested AMC perform analytic studies as a means to settle their differences throughout 1947 and 1948.

Air Staff officers sought to negotiate B-52 characteristics with little consideration or understanding of the technical compromises that new characteristics would make necessary. For instance, Maxwell argued for parasite fighters in heavy bombers. LeMay and Crawford accepted that requirement despite the performance penalties that would result, including higher aircraft weight, the need to design new items (e.g., a retrieval hook), and configuration changes in the bomb bay.

The key to understanding the conflict between AMC and the Air Staff is in the role of knowledge and analysis in decision making. The tasks and questions considered by AMC officers were relatively well-defined technically, and they preferred a strategy of incremental improvement over existing aircraft. Experimental procedures, standards of proof, and a body of background knowledge structured AMC investigations. The tasks faced by the assistant chiefs of Air Staff, in contrast, were ill-defined and most had political overtones. The challenges of design and construction were formidable and no one agreed about the ability of power plants, which existed only on paper, to meet the speed and range requirements. No power plant-propeller combination then available could cruise at speeds above 400 MPH at 35,000-feet altitude and carry a 10,000-pound payload for a 5,000-mile mission radius.

The difficulties in achieving the technical goals led to a continual search for other means to accomplish the mission. This search manifested itself in frequent alteration of the military characteristics and Boeing's provision of new or modified designs to meet those characteristics. Between November 1945 and December 1947, the heavy bomber's military characteristics were changed at least four times. Some suggestions for alterations were abandoned. Maxwell, for example, at various times proposed a parasite fighter, a small, long-range bomber without bomb bay, and weapons carried externally on pods. By the end of 1947, the desired range had been reduced from more than 13,000 miles to 8,000 miles. Aerial refueling, now available, was accepted as a means to help achieve the lower range. A small crew and structural aircraft alterations were accepted (for instance, elimination of armor, armament, and relief crews) to reduce aircraft weight and extend range.

Doubts about whether the overall airframe-engine combination would produce a satisfactory airplane led to the search for alternatives in several different directions. Air Staff officers proposed technically premature, yet seductive, ideas for heavy bomber configuration-delta wing and all-wing designs. Colonel Warden, AMC's Bombardment Branch chief, searched in another direction. He had argued in December to a doubting audience at Air Staff that turbojets were more feasible for heavy bomber propulsion than turboprops. He unofficially urged Pratt & Whitney engineers and managers-without the knowledge of Air Staff officers-to work on a turbine capable of combining with propellers or becoming a pure turbojet.

Turbojets and Swept Wings:
Uncertainty and Imperfect Knowledge

While uncertainty refers to one's inability to assign probabilities to various potential or future outcomes of actions, imperfect knowledge refers to the absence of information about outcomes that have occurred. Uncertainty and imperfect knowledge were present in different proportions and amounts in deliberations at AMC and the Air Staff. The disagreements concerning military characteristics between AMC and Air Staff officers highlighted those topics where some key Air Staff officers were more ignorant about the state of available aerodynamic knowledge. The desire of AMC officers to continue the Boeing program led to the search for means to remedy gaps in information and to reduce uncertainty about propulsion and aircraft performance.

The military characteristics generated at the end of 1947 created new technical problems for AMC. Doubts surfaced regarding whether the bomb bay would be suitable for the larger atomic weapons of the future. The AEC indicated a 12-inch increase in diameter plus increased length would be dictated by aerodynamic and ballistic characteristics of the bomb. Thus, the bomb might weigh almost 5,000 pounds more than the 10,000-pound figure given in the B-52 military characteristics. An aircraft configuration change mandated by a larger bomb would have cascading effects in the development schedule, budgets, and planning. Pending comments from Air Staff, Boeing conducted preliminary design studies and Brig Gen Donald L. Putt, the new acting assistant DCS/M, urged quick agreement between the Air Force and AEC about the dimensions and weight of atomic bombs. [96]

In the meantime, after consultations with Boeing executives, Symington ordered the Boeing contract reinstated on 26 January, and Air Staff officers favoring the B-52 cancellation searched for allies. But, there were no obvious alternatives to the B-52. While the Northrop Flying Wing showed great promise, the Air Force did not have sufficient experimental test data "to warrant placing sole reliance on this configuration for the long range bomber." Putt noted the one vital criterion in the choice of a bomber-its combat service date. He described two options: if time were critical- continue work on the B-52; if time were not critical-cancel the B-52 and hold a design competition to ensure a more advanced design. The Air Staff would have to indicate when the long-range bomber must be in service before a final program decision could be made. [97]

In a 9 February letter, Craig asked Gen Joseph T. McNarney whether he had seen a copy of Symington's letter to Boeing President Allen reinstating the B-52 contract. Given the restoration of the Boeing contract and Northrop's claims for the Flying Wing, Craig asked how to proceed with heavy bomber development and wanted an answer by 13 February when Symington was to return to Washington. [98]

Several detailed memoranda regarding the "B-52 problem" circulated at the Air Staff following Craig's letter. Craigie described the developmental history of the B-52 and concluded that the heavy bomber program was stalled despite Symington's 26 January decision to reinstate the Boeing contract. A concerted effort was needed to get the program back on "track." [99] Putt wrote that the B-52 project was unsettled because of high-level officers' desire to develop the optimum aircraft and their doubts that the current version, Model 464-29, represented the best possible design. He believed that the delays would not necessarily have adverse consequences for the Air Force, noting that the longer the Air Force waited to commit itself to an actual aircraft design, the further advanced technically that design would be. The heavy bombardment project could be stopped any time before its completion, but a new and more advanced design would entail a new (and later) delivery schedule. [100]

At AMC, senior officers attempted to save the B-52. Maj Gen Franklin O. Carroll, AMC's director of R&D, analyzed Northrop's claims of superiority for the Flying Wing, and found them wanting. The basic premise for proponents of the all-wing aircraft was that the space requirements for military stores matched the space available in the optimum wing. Under this assumption, the all-wing aircraft would be more efficient than the conventional airplane. Carroll, however, argued that Northrop seriously underestimated the space needed for military stores. More space would be needed in the aircraft, and adding a body or nacelle to contain the extra military stores would vitiate the theoretical advantages of the all-wing design. [101]

The YB-49 Flying Wing also demonstrated longitudinal instability at high speed. Little was known about this instability and it could present severe engineering difficulties. The flying wing would not be versatile in a tactical setting and would be overly sensitive to changes of center of gravity caused by the position or absence of cargo. Such problems seemed not to justify reliance on the all-wing design. Carroll concluded by recommending the conventional Boeing design and that the B-52 be accorded the highest support from Air Staff. [102]

Boeing President Allen arranged to meet with Symington on 14 February 1948. Edward C. Wells, Boeing chief engineer, presented the company's opposition to a new competition and a proposal for an aircraft weighing approximately 300,000 pounds with a range of about 8,000 miles and a speed of 500 MPH. This version was essentially the same approved by Irvine, Holtoner, Craigie, and the Aircraft and Weapons Board in November 1947.

Several days after the Symington-Allen meeting, Craig, Frederic Smith, and Craigie decided that "if the B-52 meets the requirements of the contract under which it is being bought, it will satisfy strategic requirements." These requirements included unrefueled range of approximately

8,000 miles and a cruising speed of 500 MPH over 4,000 miles of enemy territory. [103] Boeing Model 464-35 (fig. 4) matched these strategic requirements, and Air Force Undersecretary Barrows confirmed the decision to retain Boeing as prime contractor of the heavy bomber rather than adopt the Flying Wing in early March. [104]

New military characteristics for heavy bombardment aircraft were circulated in early March, with a required range of 8,000 miles. Some changes were made in required speed and altitude from the requirements in effect since December 1947. Required tactical operating altitude was increased to 40,000 feet, with 45,000 feet desired; the required speed became 500+ MPH, with 550 MPH desired. [105] These military characteristics were explained by Frederic Smith in a note to Norstad. Norstad's senior staff in DCS/O (e.g., Partridge) opposed Craig's decision to change order the B-52 rather than open a new competition. The new military characteristics, Smith wrote, "were developed as a hedge against the possible failure of the new XB-52 under development-by-the Boeing Aircraft Company." [106]

There were good reasons behind the opposition to Boeing's Model 464-35. Despite almost two and a half years of proposals, design studies, and R&D, technical uncertainties remained involving the configuration, fire control system, landing gear, and engine nacelle design. Most design changes led to unfortunate trade-offs. When Wright engineers proposed to lengthen the engine nacelles by 26 inches, range was reduced by 355 miles and cruising speed dropped three miles per hour. Boeing engineers asked Wright to study other ways to shorten the nacelle, but a solution was not immediately obvious. [107]

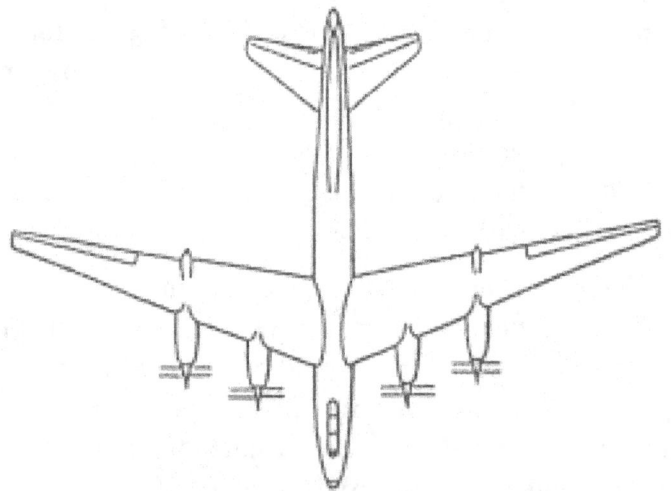

Figure 4. Boeing Model 464-35

Uncooperative relations among firms exacerbated the difficulties in finding solutions to technical trade-offs. The coordination problems could be traced to the difficulty of setting parameters and solving the technical tasks, rather than to ill will. [108] Much of the discussion about the physical size of engines, for example, was based on calculations from paper studies, an unreliable method of estimation. Yet, Wright was reluctant to give Boeing a mock-up T-35-3 with dual-rotation propeller gearing despite terms of a contract which called for such a delivery. [109]

Achieving the required range continued to be a concern. Boeing and AMC collaborated on ways to extend range. Col George F. Smith, chief of AMC's Aircraft Projects Section, suggested the incorporation of external-wing fuel cells to increase range. [110] But by June, doubts about the Model 464-35 configuration were hardening on the Air Staff. Brig Gen Thomas S. Power noted that the Perkins Committee (in November 1947) had recommended developing the XB-52 with minimum crew, range slightly above 8,000 miles, speed of 550 MPH, inflight refueling provisions, and tail armament only. Power added that, on the basis of RAND and AMC studies, this configuration would come closest to providing an atomic attack plan~ capable of operating from North American bases. These characteristics were approved by the Aircraft and Weapons Board in January 1948 over the objections of Gen George C. Kenney, commanding general of SAC, and Gen George E. Stratmeyer, commanding general of Air Defense Command. [111]

Air Staff officers were still wary about important aspects of the B-52 configuration. The B-52's weight, for instance, was a very sensitive issue as revealed by an exchange between Craigie at Air Staff and George Smith at AMC. Craigie noted that the USAF Aircraft Characteristics Summary, dated 15 May 1948 and prepared by AMC, showed the B-52 takeoff weight as 360,000 pounds. Yet, Boeing engineer Ed Wells presented to DCS/M a "finalized" B-52 proposal indicating the plane would weigh between 285,000 and 300,000 pounds. Craigie wanted to know whose figures should be believed. [112] Smith responded that the B-52 weight figure presented in the USAF Aircraft Characteristics Summary was incorrect. The summary was being revised and the correct figure would be 280.000 pounds. [113] Another example came in early October. Norstad's DCS/O staff recommended specific weight reductions of more than 9,000 pounds. Two further recommendations involved adding the capability to carry conventional bombs and reducing the maximum bomb load from 15,000 to 10,000 pounds. [114]

There were also doubts about Boeing's ability to deliver an acceptable airplane. Partridge noted that many concessions and compromises were made to speed development of the B-52. Despite "these concessions and compromises, the Air Force is receiving the B-52 the Boeing Aircraft Company wants us to have rather than the B-52 we want Boeing to build." Partridge argued, further, that if one examined the various Boeing proposals in chronological order, the Model 464-35 was too similar to Model 464 which had been rejected, the Air Force in 1946. Indeed, he noted the similarity of Models 464 and 464-35 "would lead us to the belief that the Boeing Company is giving us the old B-52 with a new coat of paint. Should this be the case, the intent of the new [military] characteristics [of March 1948] will have been defeated, and in addition, it would appear that Boeing secured the new contract, without competition, on the basis of unattainable performance figures." [115]

The occasion of mock-up inspections offered opportunities take stock of the program. Shortly before the mock-up inspection of Model 464-35 was to be held at Wright Field, Craig-the DCS/M-reviewed the state of aerodynamic and aeronautic technology with respect to the B-52. He argued .at Model 464-35 represented "the only presently known means of achieving characteristics of both long range and high speed." Indeed, he added that due to difficult performance requirements, "aerodynamic and structural margins-have been stretched farther than has been the case past designs. Growth through a series of models similar to that of the B-29 is not visualized." Range improvement, if there would be any, would come in improvements in fuel economy and by aerial refueling. [116]

Craig was both pessimistic and wrong about the state of technology. He concluded that large improvements in the heavy bombardment class of aircraft would require much new knowledge. Unless supersonic propellers became a reality, future large bombers would be powered by turbojets. "However," he continued, "neither of these developments are sufficiently near at hand that the turboprop step can be eliminated." [117]

Since Warden's initial advocacy of a turbojet design at the December 1947 B-52 Conference in General Craigie's office, doubts about the wisdom of spending large sums of money to produce a turboprop bomber, which might soon become obsolete, had been increasing at the Air Staff. The slow development and great mechanical complexity of Wright T-35 turboprop power plants had become a major source of uncertainty. Warden's position to abandon turboprop propulsion for heavy bombers was based partly on the growing complexities of this type of power source. Moreover, difficulties in T-35 development and suggestions that a dual-rotation propeller be substituted for a single-rotation propeller were causing delays in the delivery of the first flyable engines.

In the face of these problems, Warden, chief of AMC's Bombardment Branch, separately encouraged Pratt & Whitney to continue work on new turbojet designs and Boeing to study a turbojet powered heavy bomber. Figure 5 shows an experimental J57 engine lowered from the bomb bay of a B-50. In particular, Colonel Warden asked Boeing's George Schairer to, "expand a preliminary study of the performance of the airplane equipped with Westinghouse J40 [turbojet] engines and to perform studies of range extension made possible by overweight refueling on both the propeller and jet airplanes." [118] Boeing's response was designated Model 464-40. Boeing's studies indicated that the basic jet configuration (Model 464-40) compared "quite favorably" with the turboprop Model 464-35 and Air Force requirements.

Figure 5. J57 Testing on a B-50

Boeing engineer A.G. Carlsen noted the reduction in range r Model 464-40 (fig. 6) would be reasonable in view of the higher performance of the jet in climb, ceiling, cruising and high speed. He added that the jet version entailed little compromise in the basic Model 464-35 turboprop airplane—the same basic airframe and wing could be used with the jet and turboprop power plants. [119] Still, neither model's predicted range was appreciably better than Convair's B-36

Peacemaker when Boeing engineers traveled to AMC for Model 464-35 mock-up inspection in October 1948 (table 5).

On Thursday, 21 October, Warden met with Boeing personnel (including Edward C. Wells, George S. Schairer, H. Withington, Vaughn Blumenthal, Art Carlsen, and Maynard Pennell) in Dayton for the Model 464-35 mock-up inspection. They also discussed the propulsion problems of e turboprop 464-35. Warden suggested they abandon the turboprop and submit a proposal for a pure turbojet aircraft. With that proposal in mind, Boeing engineers retired to the Van Cleve Hotel to think. On Friday morning they telephoned Warden to say a proposal would be ready on Monday, 25 October. [120]

Figure 6. Boeing Model 464-40

Table 5

**Boeing Company Comparison of Basic Turboprop
and Basic Turbojet Models**

Characteristics	Turboprop 464-35	Turbojet 464-40
Range with 15,000 pound bombs and high speed run of 4,000 miles in the target region	8,000 miles	6,750 miles
High speed at 35,000 feet altitude in target range	500 MPH	536 MPH
Cruise speed for maximum range at 35,000 feet altitude and above	464 MPH	483 MPH
Service ceiling at target weight	42,000 feet	45,200 feet
Weight of external fuel and tankage	0	13,000 pounds
Gross weight after refueling	280,000 pounds	280,000 pounds
Source: Walter Boyne, Boeing B-52: A Documentary History (New York: Jane's Publishing Inc., 1982), 52.		

By October, new technologies facilitating operations, the growth of aerodynamic knowledge about swept wings, and continued development difficulties for turboprop power plants had converged to make a new type of heavy bomber possible and desirable. Among these factors were:

• The technique of inflight aerial refueling was improving; the Boeing-designed "flying boom" promised to replace and improve the efficiency of the British grapnel hook method.

• The aerodynamic drag of the swept-wing XB-47 had proved to be much lower than anticipated and much had been learned about the aerodynamics of high-speed flight.

• There was great enthusiasm for the XB-47 in both the Air Force and Boeing. A contract for the delivery of 10 B-47As had been signed on 3 September 1948.

• There was still great disagreement between Wright and the propeller manufacturers about the characteristics of a successful power plant.

The Boeing engineers submitted a balsa wood model and a 33-page proposal to Warden on Monday morning. The proposal described a large airplane with 35 degrees wing sweep, powered by eight Pratt & Whitney J57 turbojet engines paired in B-47 type pods, and a slim, low-drag angular fuselage. This latest design, Model 464-49 (fig. 7), integrated new engines with a new airframe and a new method of aerial refueling. Warden was pleased with the design. Acting on his own authority and the hope that his superiors at AMC and the Air Staff would support his decisions, Warden authorized Boeing to terminate Model 464-35 and begin work on 464-49.121 Warden promised Boeing personnel that he would deliver funding in a few months and he did.

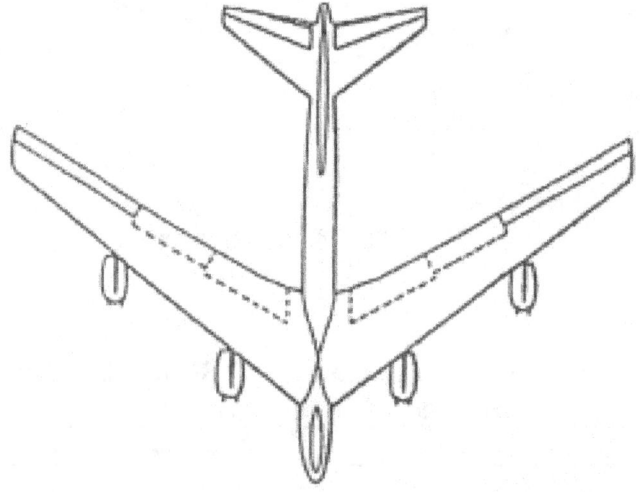

Figure 7. Boeing Model 464-49

The B-52 Development in Perspective: The Case against "Rational Management"

Between late 1945 and early 1949, the B-52's developmental history reveals a wide range of trade-offs, confusing and complex technical issues, and severe disagreements on what type of aircraft the Air Force should buy. There were at least six broad disagreements about the B-52 and the way it was being developed.

• Whether the Air Force was buying a useful capability by insisting on an aircraft with a 5,000-mile radius.

• How much a 5,000-mile-radius aircraft should weigh.

• Which platform and wing shape—straight, delta, all, or swept wing-would perform best.

• The best form of propulsion.

• How much speed and armament should be sacrificed to achieve an unrefueled 5,000-mile radius.

• The cost of the airplane and of complementary facilities, such as bases.

In addition to these disagreements, the problem of developing a heavy bomber was complicated by the interplay of multiple preferences and perceptions among officers active in the program, the movement of influential officers into or out of participation in deliberations (e.g., LeMay's departure to Europe), the sudden appearance and disappearance of various technical fixes (e.g., parasite fighters or external weapon pods), competition with the Navy for missions and budget authority, and the technical uncertainties in using new atomic weapon technology.

The B-52 design emerged from bargaining and negotiation among many officers from different offices. The contributions of AMC program managers, especially Pete Warden of AMC's Bombardment Branch, were particularly important to the success of the B-52 program. These officers were able to exert useful influence because AMC was relatively independent of the Air Staff. But the concerns of Air Staff officers provided a critical impetus to Warden's search. It seems likely that in the absence of Air Staff officers' expressed concern for military characteristics (and unexpressed competition with the Navy) Boeing and the Air Force would have settled on a less satisfactory design.

It is most significant that the exchange and interplay between the Air Staff and AMC did not conform to the Air Force's own evolving conception of a rationally organized development program. Post-World War II Air Force civilian and military leaders attempted to organize both people and programs through a "rationalization" of management. High-level officers had complained that jurisdictions, responsibilities, and communication were somewhat confused luring the war. They wanted a more rationalized and coherent organizational structure, modeled after a single, rational decision maker, with stricter delineation of tasks and "responsibilities and a reduction of administrative duplication and overlap. Within two months of the end of World War II, Assistant Secretary of War for Air Robert A. Lovett urged General Arnold to improve management techniques of the Army Air Forces. During the war, Lovett noted, Air Force officers had adopted certain "business principles to military needs and the handling of problems that are essentially those of a business enterprise," but there was still room for improvement. [122]

Conscious adoption of business methods was good public elations; it could demonstrate to the public that the Air Force was a "modern" organization. In the late 1940s, business executives thought that the fundamental rules for structuring all organizations had been discovered-rules such as chain of command, span of control, delegation of authority, specialization." [123] Top Air Force leaders believed that these fundamental rules" could be applied effectively in their organizations. Lovett's suggestion enjoyed a warm reception by Arnold and his chief deputies. [124]

From the outset of Air Force independence, the civilian leaders of the Air Force, Secretary W. Stuart Symington and assistant Secretary for Management Eugene M. Zuckert, tried to install an "effective system of management control throughout the Air Force." [125] Symington had business experience before entering government service and, from the beginning of his term as Air Force secretary, was interested in efficiency and conservation of resources throughout the department. [126] The Air Force pioneered many management tools and forms of organization. In 1946, for example, the Air Force was the first service to set up an office of comptroller, a fairly "recent development taken over from private enterprise." [127] The comptroller assumed the functions of program monitoring, statistical control, and budget and fiscal matters. Proponents of this reorganization promised economic savings and better information for high-level planning. [128] The Air Force also began to use electronic computers in 1950, before the Navy and Army, "to help solve its planning, programming, and operating problems." [129]

The application of "business principles" to the planning and conduct of R&D began in the immediate postwar years. Arnold and others argued that modern science and technology made America vulnerable to devastating attacks. [130] After seeing the value of advanced technology weapons to the war effort, Air Force leaders believed that only a long-range program of R&D could ensure that military systems would keep pace with aeronautical and missile technology. [131] The von Kármán report, Toward New Horizon, advocated a heavy reliance on science and technology in Air Force plans for new and improved weapons. The Air Force tried to respond to these officers and to the von Kármán report by grafting an R&D organization-the DC/AS for R&D-onto its traditional organization without making fundamental changes in the way R&D was conceived or managed. [132] The establishment of the DC/AS for R&D was seen, in January 1946, as part of an attempt to implement Toward New Horizons and create new channels of communication with AEC about atomic energy. [133]

Air Staff officers also sought a more effective organization to administer and manage research. This effort was partly related to the political problem of securing funding for R&D. Symbols of economy and efficiency were useful in budget presentations to Congress, and top officers' belief in the inherent value of efficiency and economy made acceptance and use of these symbols easier. Air Force military leaders thought the success of World War II production programs held promise that rational management would guide R&D effectively. Centralization of effort-the reduction of duplication and overlap of all kinds-was the centerpiece of rational management. Stanley and Weaver note that centralization of effort was "the one common theme enunciated by the various groups which sought to identify the proper role of research and development in the performance of Air Force mission objectives." [134]

Air Force officers were aided in their attempt to design a rational organization by the absence of a congressional charter for Air Force composition and organization. Since the 1947 National Security Act was vague on the subject of Air Force organization, Air Force leaders-Spaatz, Vandenberg, Norstad, and William F. McKee-"created a streamlined, functional structure" that was eventually imitated by the Army and Navy. [135] The structure of the Air Staff "was based largely on considerations other than the traditional military ones. Rapid technological advances, the greatly expanded role of logistics, and the increasing emphasis on fiscal management all combined to demand new and highly rationalized forms of organization." [136]

Management tools to secure rational control and streamlined organization had an aura of effectiveness, because they symbolized the attempt to act decisively-and evidence to establish the success or failure of these methods was not collected systematically. In fact, the bureaucratic competition between AMC and the Air Staff created a small multiorganizational system concerning questions of aircraft design. This "messier" approach was an unanticipated result of the relative independence of AMC from the Air Staff. The Air Force experience with managing B-52 R&D was one which did not rely upon precepts of what Leonard D. Sayles calls "rational advance planning" for its success. [137] The Air Force management of B-52 R&D was a dynamic process in which organizational goals-military characteristics-changed over time due to the interplay and Interaction of AMC, the Air Staff, Boeing, Wright, RAND, and Northrop.

The critics of Air Force R&D In the late 1940s (e.g., von Kármán) complained about both the clumsiness of analysis and the need for a more scientific approach to aerodynamics among Air Force leaders. A relatively atheoretical approach to aerodynamics and aeronautics by high-level officers was inherited by the postwar Air Force, especially the Air Staff, from the prewar organization. AMC officers were not always correct in their assessments, but their approach to the role of analysis and knowledge was more rigorous and cautious. Air Staff officers did have a greater interest in technological innovation in the postwar era than they had previously. However, this Interest was offset by these officers' relative Inability to understand technical issues and the increasing complexity of those issues. There was less deliberate and rigorous analysis of technical issues at Air Staff than at AMC. These different styles of analysis led to bureaucratic competition between Air Staff and AMC officers. This competition became a positive factor: it prevented the Air Force from adopting a "one-best-way" approach to heavy bomber development.

The uncertainties which AMC officers faced could not be resolved easily. Their advice on technical matters was equivocal. There were many disagreements among RAND, AMC, and Boeing engineers on a variety of issues, including weight-range estimates, wing shape, and propulsion. The heavy emphasis on turboprop development in the period 1946-1948 was an attempt to avoid the failure to pursue a promising technology. Yet, this investment created incentives to accept what turned out to be an unpromising technology. Unambiguous evidence in favor of a particular form of propulsion for heavy bombers was not available in the late 1940s. [138]

In contrast to AMC officers' use of knowledge and analysis to reduce technical uncertainty, Air Staff officers used knowledge and analysis primarily to keep the heavy bomber option undecided until a design could be discovered that would deal with political problems. For instance, RAND reports and studies projecting a high weight for the B-52 and a low probability of meeting its range requirements were used uncritically to argue for a new heavy bomber competition. The faulty premises of these RAND reports were identified by AMC officers but not by Air Staff officers. The veracity of technical claims made in the RAND reports was not the major issue in the arguments made by Air Staff officers. A larger goal was paramount: to win support for an alternate heavy bomber proposal, such as Northrop's YB-49 Flying Wing.

The description of a policy problem entails both an explanation and prescription. The B-52 development process involved disagreement regarding requirements, ambiguous and uncertain information about technology, and internal and external political conflict as senior Air Force officers sought to create an independent service and to safeguard its status and missions. Air Staff and AMC officers competed to have their views of the program accepted. Airframe firms competed with one another for a contract, and rushed to include the most recent aerodynamic knowledge in their designs. RAND, a new semi-independent analytical firm, competed with Air Force officers at AMC for status as top purveyor of quality technical advice.

The formal interaction of these different agencies on heavy bomber development allowed the Air Force, as a whole, to avoid some common disabling behaviors of organizations seeking to innovate. First, interaction prevented the Air Staff and AMC from resolving technical uncertainties by simple agreement or contract with a particular manufacturer. As the Air Staff and AMC each marshaled allies to their point of view, their mutual interaction forced them to consider trade-offs and interactions that they would have otherwise avoided. Although, at anyone time, the Air Staff and AMC advanced flawed arguments, the process of debate and exchange exposed those flaws and led to a more satisfactory decision.

Second, the interaction of the Air Staff and AMC prevented both from attending only to those familiar aspects of their environments that had produced satisfactory results in the past. AMC was skilled at creating processes to produce large numbers of aircraft; such a process required a firm commitment to a particular design. Yet, the Air Staffs constant reevaluation of military characteristics and requirements precluded premature closure on a straight-wing, propeller-driven aircraft. For the Air Staff, its ability to deal successfully with political challenges atrophied its readiness to understand technical trade-offs or to create a process to evaluate such trade-offs. Here, AMC's criticism of the Air Staffs proposals, such as for a flying wing or parasite fighters, was necessary to thwart the abandonment of the B-52 and to prevent the adoption of a technically premature idea.

Third, the interaction and mutual criticism of the complex of organizations active in the B-52 program-the airframe firms, engine and propeller firms, RAND, and the various Air Force offices-militated against the effects of "uncertainty absorption," that is, the successive editing of information as it moves up an organizational hierarchy. Such editing results in the gradual loss of the evidentiary basis of proposals or appraisals, as when Air Staff officers drew unwarranted inferences from RAND studies of the B-52's range. In many cases, mutual criticism restored that evidence and impeded action based on ill-considered ideas.

Effective innovation emerges from institutional and organizational arrangements permitting disagreement, competition, and interaction. In the development of the B-52 (fig. 8), the efforts of independent offices and firms were coordinated by the need to respond to political and substantive issues and trade-offs entailed by the task of developing an advanced aircraft. Two broad prescriptions arise from this history: tolerate duplication and overlap of policy jurisdictions, and encourage the formation of multiorganizational policy systems comprised of independent and competing organizations.

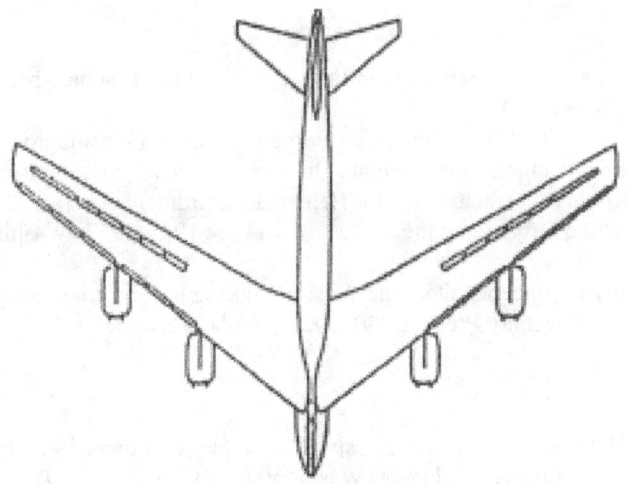

Figure 8. Boeing XB-52

Notes

1. Since the Air Force did not become a separate military service until 18 September 1947, "Air Force" used here will also refer to the US Air Corps and Army Air Forces.

2. Woodford Agee Heflin, ed., The United States Air Force Dictionary (Washington, D.C.: GPO. 1956), 422.

3. Brig Gen Alfred R. Maxwell, chief, Requirements Division, AC/AS 3, memorandum to Maj Gen E. M. Powers, AC/AS-4, subject: Military Characteristics for Heavy Bombardment Aircraft, 23 November 1945.

4. Herman Wolk, Planning and Organizing the Postwar Air Force 1943-1947 (Washington, D.C.: GPO. 1984).

5. Ibid., 36.

6. Robert Frank Futrell, Ideas. Concepts. Doctrine: Basic Thinking in the United States Air Force, vol. 1, 1907-1960 (Maxwell AFB, Ala.: Air University Press, 1989), 203-4; Wolk. 44.

7. Futrell, 215-16.

8. Ibid., 216.

9. Ibid.

10. John T. Greenwood, "The Emergence of the Postwar Strategic Air Force, 1945-1953," in Air Power and Warfare, ed. Alfred F. Hurley and Robert C. Ehrhart (Washington. D.C.: GPO, 1979), 216.

11. William H. McNeill, The Pursuit of Power: Technology, Armed Force, and Society since A.D. 1000 (Chicago: University of Chicago Press, 1982), 355.

12. Futrell, 216.

13. Trevor Gardner, "How We Fell Behind in Guided Missiles." The Air Power Historian 5: 2-13, no. 1, January 1958, 8. Gardner was assistant secretary of the Air Force for research and development.

14. David Alan Rosenberg, "American Postwar Air Doctrine and Organization: The Navy Experience," in Air Power and Warfare, ed. Alfred F. Hurley and Robert C. Ehrhart (Washington, D.C.: GPO, 1979), 247.

15. Greenwood, 227.

16. David Alan Rosenberg, "U.S. Nuclear Stockpile, 1945 to 1950," The Bulletin of the Atomic Scientists, May 1982, 27.

17. Ibid., 27.

18. Ibid., 28-29.

19. Rosenberg, "American Postwar Air Doctrine," 249.

20. Ibid., 250.

21. Ibid., 253.

22. Heflin, 423.

23. Rosenberg, "American Postwar Air Doctrine," 254.

24. Gordon Swanborough and Peter M. Bowers, United States Navy Aircraft Since 1911, 2d ed. (Annapolis, Md.: Naval Institute Press. 1976), 186.

25. Ibid., 186-88.

26. Briefing, Air Materiel Command, subject: XB-52 [January] 1949.

27. Rosenberg, "American Postwar Air Doctrine," 255.

28. Col George E. Price, chief, Aircraft Projects Section, Engineering Division, Air Technical Service Command, memorandum to Boeing, subject: Proposal for Heavy Bombardment Airplane, 13 February 1946.

29. Margaret Bagwell, The XB-52 Airplane (Wright-Patterson Air Force Base [AFB]: Historical Office, Air Materiel Command, August 1949).

30. Bombardment Branch. Engineering Division, AMC, telegram to George Schairer and L. A. Wood. Boeing. 25 April 1946.

31. William M. Allen, president. Boeing Aircraft Company to Lt Gen Nathan F. Twining, commanding general, AMC, letter, subject: Cost Plus Fixed Fee Proposal for Heavy Bombardment Airplane, Boeing Aircraft Company Model 462, Phase I, 18 April 1946.

32. Brig Gen Laurence C. Craigie, chief, Engineering Division, AMC to Gen Carl A. Spaatz, commanding general AAF; attn.: AC/AS-4, letter, 23 May 1946.

33. Col John G. Moore, deputy AC/AS-4, memorandum to Lt Gen Nathan F. Twining, commanding general AMC, 29 May 1946.

34. H. S. Lippman, Aircraft Projects Section. Engineering Division, AMC, memorandum report, subject: Comments on Model Specification, MAC, Model 462 (XB-52) Airplane, no date (the report was requested on 5 June 1946).

35. Col George E. Price, chief, Aircraft Projects Section, Engineering Division, AMC memorandum to Design Branch, Aircraft Laboratory, Engineering Division, AMC; attn.: Maj J. F. Wadsworth, subject: Conversations Regarding "Heavy Bomber Studies," 13 August 1946.

36. Maj Gen E. M. Powers, AC/AS-4, to AC/AS-3, Operations and Training, routing and record sheet, attn.: Brig Gen Alfred R Maxwell, subject: Presentation of 1948 Research and Development Budget, 19 September 1946.

37. Bagwell, 22-23.

38. Ibid., 19.

39. Brig Gen Alfred R Maxwell, chief, Requirements Division, AC/AS-3, memorandum to Maj Gen Earle E. Partridge, AC/AS-3, subject: XB-52 Contract, 27 November 1946.

40. Brig Gen Laurence C. Craigie, chief, Engineering Division, AMC, to Gen Carl A. Spaatz, commanding general, AAF, attn.: AC/AS-4, letter, subject: Notes of Conference at Wright Field with AC/AS-3 Personnel on XB-51, XB-52, XB-53 and Military Characteristics in General, 26 November 1946.

41. Maxwell.

42. Ibid.

43. Maj Gen E. M. Powers, AC/AS-4, memorandum to Lt Gen Nathan F. Twining, commanding general, AMC, subject: XB-52, 7 December 1946,

44. Report of Conference, subject: Discussion of Reduction Gear Ratio of T-35 WAC Turbine, 20 August 1946.

45. Ibid.

46. Robert Schlaifer. Development of Aircraft Engines (Boston: Graduate School of Business Administration. Harvard University. 1950), 445.

47. Thomas A. Marschak, "The Role of Project Histories in the Study of R&D," in Econometrics and Operations Research, vol. 8, Strategy for R&D Studies in the Microeconomics of Development, ed. Thomas A. Marschak, Thomas K. Glennan Jr. and Robert Summers (New York: Springer-Verlag, 1967), 83.

48. Ibid., 84.

49. George S. Schairer, staff engineer, Aerodynamics and Power Plant, Boeing, to Carl F. Baker, chief engineer, Hamilton-Standard Propeller, Division of United Aircraft Corporation, letter, subject: Propeller Design Information for XB-52 Airplane, 13 December 1946.

50. Maj William C. Brady, Research and Engineering Division, Assistant Chief of Air Staff-4 (AC/AS-4), memorandum for record, subject: XB-52 Conference held 7 January 1947.

51. Ibid.

52. Col George Smith, Bombardment Branch, Aircraft Projects Section. Engineering Division, AMC, routing and record sheet, to Aircraft Laboratory, Engineering Division, AMC; attn., Col John G. Moore, subject: Landing Gear Requirement for Very Large Airplane, 13 January 1947; Col John G. Moore, Aircraft Laboratory, Engineering Division. AMC, routing and record sheet to Col George Smith. Bombardment Branch, Aircraft Projects Section, Engineering Division, AMC, subject: Reply to Routing and Record Sheet, 13 January 1947, re: "Landing Gear Requirement for Very Large Airplane," 18 February 1947.

53. William M. Allen, president, Boeing, to Lt Gen Nathan F. Twining, commanding general AMC, attn.: TSESA-2, Bombardment Branch, Aircraft Project Section, Engineering Division, letter, subject: Contract W33-038 ac-15065, AAF Model XB-52 Airplane (Boeing Model 464-17), 19 February 1947.

54. Brig Gen S. R. Brentnall, Engineering Operations, Engineering Division, AMC, memorandum to Gen Carl A. Spaatz, commanding general AAF, attn.: AC/AS-4, subject: Contract W33-038 ac-15065, Continuation of the XB-52 Project, 17 April 1947.

55. Brig Gen Alden R. Crawford, chief. Research and Engineering Division, AC/AS-4, memorandum to Gen Nathan F. Twining, commanding general, AMC, 2 May 1947.

56. Brig Gen S. R. Brentnall, Engineering Operations, Engineering Division, AMC, memorandum to Gen Carl A. Spaatz, commanding general AAF, attn.: AC/AS-4, 2 June 1947; transcript of telephone conversation, Alden R Crawford. George Smith. Donald Putt. 14 May 1947; Brig Gen Alden R Crawford, chief, Research and Engineering Division, AC/AS-4, memorandum to Gen Nathan F. Twining, commanding general AMC, attn.: TSTEX (TSEOA-2), 16 June 1947.

57. Maj Gen Earle E. Partridge. AC/AS 3, memorandum to Maj Gen Curtis E. LeMay, DC/AS for R&D, subject: Defensive Armament in Bombardment Aircraft, 5 March 1947.

58. Brig Gen Alfred R. Maxwell, chief, Requirements Division, AC/AS-3, memorandum to Brig Gen Alden R Crawford, Research and Engineering Division, Office, AC/AS-4, subject: XB-52 Performance, 21 April 1947.

59. Ibid.

60. Maj Gen E. M. Powers. AC/AS-4 to Lt Gen Nathan F. Twining, commanding general, AMC, attn.: Brig Gen Laurence C. Craigie. Engineering Division, AMC, letter, subject: B-52, 25 April 1947.

61. Ibid.

62. Maj Gen Curtis E. LeMay, DC/AS for R&D, to Lt Gen Nathan F. Twining, commanding general, AMC, letter, 15 May 1947.

63. Ibid.

64. Maj Gen E. M. Powers, AC/AS-4, to Lt Gen Nathan F. Twining, commanding general, AMC, attn.: Engineering Division, AMC. Letter, subject: Medium Bombardment Aircraft, 8 May 1947.

65. LeMay.

66. J. A. Boykin, Aircraft Project Section, AMC, memorandum report, subject, Analysis of the XB-52 Project, 23 June 1947.

67. Ibid.

68. Brig Gen Alfred R. Maxwell, chief, Requirements Division, AC/AS-3, memorandum to Maj Gen E. M. Powers, AC/AS-4, subject: Military Characteristics of Aircraft, 23 June 1947.

69. Maj Gen Laurence C. Craigie, chief. Engineering Division, AMC, to Gen Carl A. Spaatz, commanding general, AAF, attn.: AC/AS-4, letter, subject: XB-52 Airplane, 11 July 1947.

70. Ibid.

71. Ibid.

72. Ibid.

73. Maj Gen Curtis E. LeMay, DC/AS for R&D to Lt Gen Nathan F. Twining, commanding general, AMC, and Maj Gen L. C. Craigie, chief, Engineering Division, AMC, letter, subject: XB-52 Program, 14 July 1947; see also, Maj Gen Curtis E. LeMay, DC/AS for R&D to Lt Gen Nathan F. Twining, commanding general, AMC, letter, subject: Earlier Letter on XB-52 Program, Possibility of Other Ways to Meet the Primary Long Range Mission, 1 August 1947.

74. LeMay to Twining, letter, 14 July 1947.

75. George S. Schairer, staff engineer, Aerodynamics and Power Plant, Boeing, to R. E. Gould, Aeroproducts Division, General Motors Corporation, letter, subject: Propeller Design Information for B-52 Airplane, 10 September 1947; Boeing to Lt Gen Nathan F. Twining, commanding general, AMC, attn.: TSEOA-2, letter, subject: Contract W33-038 ac-15065, Transmittal of Preliminary Design Data, 7 August 1947.

76. Col George F. Smith, chief, Aircraft Projects Section, Engineering Division, AMC, to Curtiss Propeller Division, Curtiss-Wright Corporation (the same letter was sent to Aeroproduct Division and Hamilton-Standard Propellers), letter, subject: Coordination of Development Projects Related to the XB-52 Airplane Project, 29 September 1947.

77. Col George F. Smith, chief, Aircraft Projects Section, Engineering Division, AMC, routing and record sheet to TSEAC, subject: Heavy Bomber Design Study, 9 October 1947.

78. Maj Gen Laurence C. Craigie, Director of Research and Development, DCS/M. routing and record sheet to Requirements Division, Deputy Chief of Staff/Operations (DCS/O), subject: XB-52. 6 November 1947.

79. Ibid.

80. Summary of Meeting, "Conference with Bombardment Subcommittee, Air Force Aircraft and Weapons Board to Discuss the Military Characteristics for Strategic Bomber Requirements," 18 November 1947.

81. Ibid.

82. Maj Gen Earle E. Partridge, director, Training and Requirements, DCS/O, routing and record sheet, to Maj Gen Laurence C. Craigie, director of research and development, DCS/M, subject: XB-52. 19 November 1947.

83. Col Clarence S. Irvine, assistant to the chief of staff Headquarters, SAC to Maj Gen L. C. Craigie, director of research and development, DCS/M, letter, 28 November 1947.

84. Ibid.

85. Col J. S. Holtoner, chief, aircraft branch, Office, DCS/M, memorandum to Maj Gen L. C. Craigie, director of research and development, DCS/M, subject: XB-52, 28 November 1947.

86. Ibid.

87. Conference notes, Bombardment Branch, Engineering Division, AMC, subject: "Conference on XB-52 with the Director of Research and Development, 1 December 1947," 7 December 1947.

88. Maj William C. Brady, chief, Bomber Section, Aircraft Branch, Deputy Chief of Staff/Materiel (DCS/MI. memorandum for record, subject: XB-52 Conference, 2 December 1947.

89. Air Staff Summary Sheet, Maj Gen Laurence C. Craigie, director of research and development, DCS/M, Heavy Bombardment Aircraft, approximately 1 December 1947; Gen Hoyt S. Vandenberg, vice chief of staff, memorandum to W. Stuart Symington, secretary of the Air Force, subject: Heavy Bombardment Aircraft, approximately 1 December 1947.

90. Maj Gen Earle E. Partridge. Director, training and requirements, DCS/O to Brig Gen Frederic H. Smith Jr., chief, Requirements Division, DCS/O, letter, subject: New Heavy Bomber Contracts, 8 January 1948.

91. William M. Allen, president, Boeing, to Secretary of the Air Force W. Stuart Symington, letter, subject: Proposed Competition for Heavy Bombardment Airplane and Contract W33-038 ac-15065 with Boeing for Phase I and Limited Phase II Program for Heavy Bombardment Airplane (Model XB-521, 26 December 1947.

92. Gen Joseph T. McNarney, commanding general, AMC, memorandum to Maj Gen Laurence C. Craigie, director of research and development, DCS/M, subject: 11 December Letter on Heavy Bombardment Aircraft, 30 December 1947.

93. Edwin P. Hartman, Adventures in Research: A History of Ames Research Center 1940-1965 (Washington. D.C.: NASA, 19701, 55.

94. George S. Schairer, 'The Role of Competition in Aeronautics," The Wilbur and Orville Wright Memorial Lecture of the Royal Aeronautical Society, London, England, 5 December 1968, 18.

95. James G. March and Herbert A Simon, Organizations (New York: Wiley, 1958), 131.

96. Brig Gen Donald L. Putt, deputy chief, Engineering Division, AMC to Gen Carl A. Spaatz, CSAF; attn.: Lt Gen Howard A Craig, DCS/M, letter, subject: Heavy Bombardment Aircraft, 26 January 1948.

97. Brig Gen Donald L. Putt, acting assistant DCS/M. memorandum to Lt Gen Howard A. Craig, DCS/M. subject: B-52 Problem, 10 February 1948.

98. Lt Gen Howard A Craig. DCS/M, to Gen Joseph T. McNarney, commanding general, AMC, letter, 9 February 1948.

99. Maj Gen Laurence C. Craigie, director of research and development, DCS/M, memorandum to Lt Gen Howard A. Craig, DCS/M, subject: Development of Heavy Bombardment Aircraft, 13 February 1948.

100. Putt, memorandum to Craig, 10 February 1948.

101. Maj Gen Franklin O. Carroll, director, research and development, Engineering Division, AMC, to Gen Carl A Spaatz, chief of staff of Air Force (CSAF), attn.: Lt Gen Howard Craig, DCS/M, letter, subject: Heavy Bombardment Configuration, 13 February 1948.

102. Ibid.

103. Lt Gen Howard A Craig, DCS/M, memorandum to Gen Hoyt S. Vandenberg, vice chief of staff (VCS), subject: Capability of B-52 to Perform Long Range Strategic Mission, 17 February 1948.

104. Undersecretary of the Air Force Arthur S. Barrows, memorandum to Secretary of the Air Force W. Stuart Symington, 5 March 1948; Maj Gen Earle E. Partridge, director, training and requirements, DCS/O, routing and record sheet to Maj Gen Laurence C. Craigie, director, research and development, DCS/M, subject: Heavy Bombardment Aircraft, 12 March 1948.

105. Col Leslie O. Peterson, chief, Requirements Division, Directorate of Training and Requirements, DCS/O, memorandum to Lt Gen Howard A Craig, DCS/M, subject: Military Characteristics for Heavy Bombardment Aircraft, 3 March 1948.

106. Maj Gen Frederic H. Smith Jr., assistant for programming, DCS/O, memorandum to Lt Gen Lauris Norstad, DCS/O, subject: Military Characteristics for New Heavy Bombardment Aircraft, 8 March 1948.

107. A G. Carlsen, Boeing, chief project engineer to Gen Joseph T. McNarney, commanding general, AMC, letter, subject: Contract; Engine Accessory Gear Box Drive Installation, 30 March 1948.

108. Aircraft Projects Section, Air Materiel Command (AMC), memorandum report, subject: Conference on Propellers for the XB-52 Airplane, 8 June 1948.

109. Col George F. Smith, chief, Aircraft Projects Section, Engineering Division, AMC, to Power Plant Laboratory, attn.: Marvin Bennett, letter, 7 April 1948.

110. Col George F. Smith, chief, Aircraft Projects Section, Engineering Division, AMC, to Boeing, letter, subject: XB-52 Airplane, Range Extension, 15 June 1948.

111. Maj Gen Earle E. Partridge, director, training and requirements. DCS/O, memorandum for record, subject: New Medium Bomber and B-52 Armament, 14 June 1948.

112. Maj Gen Laurence C. Craigie, director of research and development, DCS/M, to Gen Joseph T. McNarney, commanding general, AMC, letter, 30 June 1948.

113. Col George F. Smith, chief, Aircraft Projects Section, Engineering Division. AMC, to Gen Hoyt S. Vandenberg, CSAF, attn.: Maj Gen Laurence C. Craigie, letter, 19 July 1948.

114. Air Staff Summary Sheet, Gen Lauris Norstad, Weight Reduction and Simplification of B-52 Airplane, 6 October 1948.

115. Maj Gen Earle E. Partridge, director, training and requirements, DCS/O, memorandum to Maj Gen Laurence C. Craigie, director, research and development, DCS/M, subject: B-52 Program. 15 June 1948.

116. Lt Gen Howard A Craig, DCS/M, memorandum to Gen Joseph T. McNarney, commanding general, AMC, subject: XB-52 Long Range Bombardment Airplane, 16 October 1948.

117. Ibid.

118. A G. Carlsen. Boeing, chief project engineer, to Gen Joseph T. McNarney, commanding general, AMC, letter, subject: Contract W33-038 ac-15065-MX-839; Model XB-52 Variant Studies-Range Extension, 28 July 1948.

119. Ibid.

120. Walter Boyne, Boeing B-52: A Documentary History (New York: Jane's Publishing Incorporated, 1982), 50; Harold Mansfield, Vision: A Saga of the Sky (New York: Duell, Sloan & Pearce, 1956), 308-9.

121. Boyne, 52.

122. Harry Borowski, "Air Force Leaders and Business Methods," paper presented to the joint meeting of American Military Institute, Military Classics Seminar, and Air Force Historical Foundation, 13-14 April 1984, Bolling AFB, D.C.

123. Ross A Webber, "Staying Organized," The Wharton Magazine 3, no. 3 (Spring 1979): 16.

124. Borowski.

125. Harold Larson, "Management," in A History of the United States Air Force, ed. Alfred Goldberg (Princeton: D. Van Nostrand Co., Inc., 1957); Alfred Goldberg, "The Worldwide Air Force," in Goldberg, 106.

126. John D. Glover and Paul R. Lawrence, A Case Study of High Level Administration in a Large Organization: The Office of the Assistant Secretary of the Air Force (Management), 1947-1952 (Cambridge: Riverside Press, 1960), 14.

127. Larson, 227.

128. Borowski.

129. Larson, 231.

130. Greenwood, 219.

131. Dennis J. Stanley and John J. Weaver, An Air Force Command for R&D, 1949-1976: The History of ARDC/AFSC (Washington, D.C.: AFSC, History Office, n.d.).

132. Donald Ralph Baucom, "Air Force Images of Research and Development and Their Reflection in Organizational Structure and Management Policies" (PhD diss., University of Oklahoma, 1976), 68-69.

133. Greenwood, 219.

134. Stanley and Weaver, 12; see also McNeill's comments on the "primacy of command over market as the preferred means for mobilizing large-scale human effort." William H. McNeill, The Pursuit of Power: Technology, Armed Force, and Society Since A.D. 1000 (Chicago: University of Chicago Press, 1982), 316-17.

135. Goldberg, "The Worldwide Air Force," 106.

136. Ibid., 110; George M. Watson, The Office of the Secretary of the Air Force 1947-1965 (Washington, D.C.: GPO, 1993), 51-52.

137. Leonard D. Sayles, "Technological Innovation and the Planning Process," Organizational Dynamics, Summer 1973, 68-80.

138. Marschak.

Chapter 6
Conclusion

Novices in mathematics, science, or engineering are forever demanding infallible, universal, mechanical methods for solving problems.

-J. R. Pierce

How is the period 1945-1948, when jet propulsion was being introduced into the Air Force, relevant to current conditions? How should senior decision makers perceive and understand past policies and organizations designed to encourage military innovations? How may their representation or understanding of the policy process be improved? And, does the adoption of jet propulsion for the B-52 provide a benchmark from which we may examine major changes in the American style of conceiving, creating, and implementing military innovations?

Novices in policy analysis, like the novices in the technical fields described by physicist /John R. Pierce, often recommend simple and infallible procedures, Yet, the analysis of military innovation early in this study demonstrates the complexity and multiple levels of factors influencing innovation. The major lesson of this study is that the creation of small stand-alone organizations dedicated to developing new technology may not encourage the desired innovative thinking. Rather, prospects to innovate new weapons and operational concepts are enhanced by improving the interaction between organizations with overlapping jurisdictions. [1] Through this multiorganizational interaction it is possible to remove (or mitigate) organizational pathologies and identify and correct errors. The introduction of jet propulsion into the B-52 illuminates the effort, time, attention, resources, expense, and good luck required to innovate in the military services, involving a complex mixture that creates many opportunities for error, schedule delays, or cost overruns. Military innovations involve questions about politics, cooperation and coordination, and social benefits, and like other development efforts, there appears to be no error-free method to predict-at the start of a particular program-the end results.

This concluding chapter reflects on the material presented in the study to (1) summarize observations about innovation and military revolutions, (2) identify and critique significant policy-relevant features of the jet propulsion B-52 case, and (3) review the argument in favor of a multiorganizational approach to exploiting rapidly advancing technology.

How Does the Pace of Scientific and Technological Advance Effect Acquisition Decisions?

The current rapid and accelerating pace of scientific and technological development presents policy makers, military organizations, and the political system with choices that are fundamentally different from earlier eras: these choices must be made in a context of great complexity. Applied to the military arena, the process of invention and incremental improvement of new technologies and capabilities across a wide range of equipment and tasks has created an interconnected, self- generating or self-reinforcing dynamic of change, where each invention or improvement leads to new requirements and solutions. The cycle overlaps and accelerates, providing further opportunities for novel products and tasks, and the concomitant extinction of superseded tasks and products. Invention, incremental change, and organizational response together act as .a positive feedback, through which actions are amplified beyond those originally anticipated, designed, or desired.

Existing weapons acquisition procedures and bureaucracies were established to create weapons with predictable effects on combat and military organization. Most individual acquisition procedures are simple. Despite simple rules, however, the number of interrelated actors and organizations acting according to those rules actually make the system very complex. This complexity sensitizes the weapons acquisition process to chance actions and initial conditions so that small differences in the starting composition of development programs may lead to drastically different outcomes responses to problems. Management functions in a context of risky organizational problems that are not under control.

For example, while coping with incomplete and imperfect knowledge, managers must respond to political demands, including concerns about costs versus delivery schedules. Such trade-offs cannot be resolved in advance and will differ in every development program. Designers of military development processes should understand that error-intolerant individuals and organizations are unlikely to identify and nurture an immature, but revolutionary complex of technologies, and operational concepts. They treat unanticipated results as errors that are inconsistent with established goals-for example, doctrine or the procurement of particular technologies. Risk-loving individuals and organizations, in contrast, search and welcome the kind of unanticipated results that magnify combat capability, but are less likely to identify technological or operational failures quickly enough.

Identifying errors in the military acquisition decision process hinges on enhancing policy makers' ability to apply knowledge and analysis to problems. The positive experience of the introduction of carrier aviation into the US Navy and jet propulsion into the US Air Force (and the negative experience of the failure of the US Army to develop the tank more fully and create an analogue to the blitzkrieg) suggests that policy analysis should be directed to improving the quality of interactions among individuals and organizations-that is, to highlight intelligent criticism of programs while resisting the tendency to allow a single organization (or individual) to make decisions by intellectual fiat. An appropriately complex structure of social and intellectual interaction will improve military acquisition over the long-run, encouraging the effective accumulation and application of useful knowledge about force structure mixtures.

Should Policy Makers Seek a Coherent Framework for Acquisition Decisions ?

Mark Twain said, "Get your facts straight, then you can distort them as you see fit." Cognitive psychologists have observed that people consistently exaggerate after an event.

Earlier military revolutions were realized in military organizations that developed weapons and identified and corrected doctrinal errors through a years-long process of trial- and-error and incremental change. The creation of new military roles and specializations appropriate to revolutionary changes involves the agreement and support (or the absence of opposition) of officers of all ranks, and such a social change takes years to accomplish. The greater social complexity of modem military organizations, coupled with the political demands to develop new systems in a shorter period of time, puts ever more stringent demands on the administration and management of innovation. Today's very complex man- machine organization systems inevitably beget novelty-we cannot stop it. In such systems we must manage-not control-self-generating technological change, understanding that goals and plans will be superseded, and a process of reasoned criticism will be necessary to identify errors, to highlight unanticipated implications, and to propose effective solutions.

What Distinguishes Managing from Controlling Technological Growth?

Maintaining a distinction between management and control is crucial to fostering innovation-military or otherwise. The concepts of management and control are not synonymous. In organizations the concept of control implies the ability to determine phenomena and events, knowledge of cause and effect relations, and associated procedures to apply that knowledge. With reliable knowledge, the "manager" need only ensure compliance with the procedures to achieve the desired outcomes. In military acquisition, however, complete, reliable, and verified cause-effect knowledge relating emerging or "innovative" technologies and operational concepts to combat outcomes does not exist in the early stages of a project, and becomes clear only with the evaluation of combat outcomes. [2] In such a situation, the concept of control is useful. The concept of management, in contrast, assumes incomplete knowledge, uncertainty, and the consequent necessity of more flexible what could have been anticipated with foresight; they also misremember their own predictions about future events to exaggerate in hindsight what they thought they knew in foresight. [3] This characteristic of human thought represents a major methodological pitfall for historians, policy makers, and analysts. The historian's knowledge of outcomes of events or processes and his desire to create a coherent narrative make it easy to exaggerate the foresight of historical actors. Such exaggeration limits the utility of historical analysis to the policy maker because it encourages the view that participants in a historical situation were fully aware of its eventual importance or impact (e.g., "Dear Diary, Today the Gutenberg press began a revolution in literacy that will lead to the industrial revolution.") and the myth of the critical experiment that unambiguously establishes the validity of one theory, technology, or operational concept. In fact, critical experiments are identified only with hindsight. [4]

The policy process is neither linear, orderly nor simple. It entails disparate streams of activity that coalesce due to the need to make a decision. The separate streams of activity in the organization are strongly influenced by the ubiquity of ambiguous evidence (unavailable, unreliable, or deceptive information), fluid participation by officials (officials devote varying amounts of energy and attention to a given issue), and conflicting preferences. The participation of officials on various issues or tasks is constrained by other demands on their time and attention, so that no single official dominates the decision process in all its phases, nor are all issues considered simultaneously. Finally, bureaucracies operate with all sorts of ill-defined or inconsistent preferences. Such inconsistent and poorly defined preferences are often concealed until the need to take action forces those holding them to speak up.

The case study presented in these pages of how jet propulsion was introduced into the B-52 shows no coherent search process inevitably leading to the choice of a particular technology. Nor was decision-making concerning the B-52 development program coherent or orderly. Different mixtures of participants, problems, and solutions came together at various times to make decisions about continued funding or to review the status of performance projections and requirements. In 1948, for example, the assignment of Gen Curtis E. LeMay to Europe and his absence from B-52 discussions was a major factor allowing Col Henry E. "Pete" Warden to secure Air Staff approval of his initial swept-wing turbojet design. This design, it should be remembered, did not meet the Air Staffs range requirement, and aerial refueling had only been approved seven months before in March 1948. Decisions regarding the status of the B-52 program depended upon the connections among the different streams of activity in the multiorganizational system of AMC, the Air Staff, and Boeing and the engine firms.

The case study illustrates the reality that surprises and failures are inevitable in development programs where information and knowledge are indeterminate, ambiguous, and imperfect. Contrary to conventional wisdom, clarity of vision is not a property of successful innovation. Advocates of "clear vision," assuming a linear historical path, overlook the large number of times partisans of a technology or operational concept have been wrong. Roberta Wohlstetter's conclusions about the surprise attack on Pearl Harbor are as relevant to technological and operational innovation as they are to intelligence failure. She wrote, "We have to accept the fact of uncertainty and learn to live with it. No magic, in code or other- wise, will provide certainty. Our plans must work without it." [5]

How Did the Newness of Jet Propulsion Effect Its Adoption by the Air Force?

An innovation must be much more beneficial than the existing procedure or technology before the "flow of benefits compensates for the relative weakness of the newer" organizational relationships mandated by the innovation. The process of inventing new roles has high costs in terms of worry, time, conflict, and temporary inefficiency. Where the "liability of newness" is great, organizational innovation will tend to be carried out only when the alternatives are bleak.

The liability of newness did little to influence the acceptance of the jet-propelled B-52, because so few organizational relationships were changed. Jet-propelled aircraft did not require new operational concepts; they used and extended existing concepts. Mechanics who serviced propeller-piston engine propulsion systems learned how to maintain and repair gas turbine power plants. The propeller firms had less influence over the acquisition and design process than did the airframe and engine firms, and so were unable to stymie the transition. By 1948 leaders of airframe and engine firms could see the higher performance of jet-powered aircraft and embraced the opportunity to produce the new aircraft.

How Did Oversight Effect the Adoption of Jet Propulsion in the B-52 Program?

Detailed oversight and review occurring too early can discourage or impede the implementation of innovation. Too many veto points in other parts of the organization can stifle an innovation before it has a chance to prove its value. Isolating the innovative group often promotes success in implementing an innovation by minimizing negotiation and transaction and coordination costs. In this vein, philosopher Michael Polanyi describes a situation where ignorance of criticism prevented the abandonment of a good idea-his theory of adsorption. Polanyi writes "I would never have conceived my theory, let alone have made a great effort to verify it, if I had been more familiar with major developments in physics that were taking place. Moreover, my initial ignorance of the powerful, false objections that were raised against my ideas protected those ideas from being nipped in the bud." [6] Nevertheless, some oversight or review is necessary to avoid other types of failures: schedule delays, cost overruns, performance shortfalls, or mismatch between requirements and operational capability.

The Air Staff oversaw, criticized, reviewed, and threatened a number of times to cancel the AMC-managed B-52 development program. Air Staff review (and negative evaluation) of Boeing's efforts began within a few months of program initiation. The oversight and criticism stimulated AMC and Boeing to search for more capable designs. The organizational context of the Air Staffs oversight was critical. It was conducted within a loosely coupled organizational structure: AMC personnel, reporting to their own hierarchical superiors, were separate and partially independent from the Air Staff. This independence fostered debate and argument between Air Staff and AMC officers, and led each to search for better information and analyses to support their respective views. As a result, several organizational pathologies were minimized, including "premature programming" (settling on a particular design before appropriate knowledge and information had been gathered) and "uncertainty absorption" (successive removal of the evidentiary basis of analysis as it moves up the organizational chain). However, while the Air Staffs review was instrumental in stimulating a capable design, it also contributed to schedule delays and some extra expenditures.

How Did Chance Effect the Adoption of Jet Propulsion?

Many critical discoveries have depended on fortuitous events that were seen many times before they were recognized. Chance events or actions effect the invention or implementation of innovations in several ways. Historical accidents sometimes enable a technology to gain an early lead over competing technologies to "corner the market" and lock potential competitors out of consideration. An established technology may become so dominant that superior alternatives developed subsequently cannot supplant it.

On the one hand, the timing of the introduction of jet propulsion into the Air Force partly depended upon chance. Gen Hap Arnold's initial contact with the technology took place because of chance encounters between a US military liaison and his British counterparts. In 1941 Arnold's interest in, and prediction of, the potential value of jet propulsion saved the US military some time in developing jet fighters. But Great Britain and Germany were working on jet aircraft, so US military leaders would have been alerted to the technology within a few years anyway.

The relatively quick integration of swept-wings with jet propulsion between 1945 and 1948 owed much to the fortuitous presence of a Boeing aerodynamicist in Germany. George Schairer Boeing's chief aerodynamicist, first heard of wing sweep from Bob Jones of NACA before World War II. But, Boeing not have the jet propulsion systems that could exploit the swept-wing airframe's speed characteristics. This situation changed in 1945 as US civilian and military teams went to Germany to evaluate the latest German technology. Schairer was a member of a team that reviewed Adolf Busemann's wind tunnel files on wing sweep. He immediately directed the Boeing team in Seattle to investigate wing sweep to meet the Air Corps' 1943 contract for a medium bomber (see chap. 4). [7] The aircraft resulting from the redirected research effort was the B-47.

Chance played an even more important role in Col Pete Warden's proposal to Boeing to create a swept-wing turbojet bomber. Warden was "at the right place, at the right time." As the chief of AMC's Bombardment Branch, he oversaw the development of long-range bombers. He had earned a master's degree in aeronautical engineering from Massachusetts Institute of Technology and could evaluate technology forecasts. He also had access to skilled civilian and military aerodynamicists (including some German engineers). Warden was able to forecast the success of a swept-wing turbojet design because, functioning at the center of overlapping technical communities, he was able to combine information about manufacturing techniques, new materials, and aerodynamics with practical experience in turbines and medium bombers.

However, it is difficult to argue that combining wing sweep with jet propulsion was entirely a result of chance circumstances that brought out their unseen potential. Improvements in manufacturing techniques, materials, and designs not available during the interwar period were brought into being because of the war effort. While Boeing engineers could not know that their design would prove so successful, they could reasonably argue that their turbojet swept-wing design solved critical technical and operational problems and opened a new and unexplored set of opportunities to execute the Air Force's strategic mission. The lesson for military policy makers may be to earmark a portion of R&D monies to produce basic engineering knowledge about materials, design concepts, and manufacturing technology that will support future technological innovation. Additional development funds could back technology demonstration projects that explore technologies that have been "seen, but not adopted." Attention to basic engineering knowledge furnishes the acquisition community with two key advantages: a growing stock of knowledge to overcome unknown future technological obstacles and a reduced (but not eliminated) role for the sort of luck that Whittle (finally) enjoyed (his early backers would not have invested in the turbojet concept had they known how many technical obstacles had to be overcome).

Do New Technologies Immediately Drive Out the Established Ones?

It takes many years to accept new technologies and associated operational concepts. In the case of jet propulsion, propeller-piston propulsion systems were not abandoned overnight; propeller-driven aircraft still remain in service. During the transitional period from one set of technologies and operational concepts to another, good arguments may be presented for continued reliance upon the existing way of doing things. [8] Innovations may be accepted within shorter periods (one or two years) if the sources of ideas are close to those agencies responsible for enactment, little time or effort need be devoted to research, and every feasible alternative is not tested.

The Air Force's embrace of the B-52 continued to be tentative after the acceptance of Boeing's swept-wing turbojet proposal; alternative aircraft and propulsion systems had advocates and supporters. [9] Nevertheless, the Air Force's overall endorsement of jet propulsion was relatively quick. Those responsible for use of jet-propelled bombers, including SAC commander General LeMay, supported the acquisition of jet aircraft. The swept-wing turbojet version of the B-52 was adopted less than a year after promising test results began emerging about the B-47 and only five years after the first flight of the Bell XP-59, the first US jet fighter.

What Factors Encourage the Exploitation of New Technology?

Exploiting new technology and the advancing rate of technological change requires the application of empirical premises and assumptions to military acquisition. In practice, an empirical stance dictates testing, hypothesis and experimentation, and trial and error. Yet. as in civilian government and private organizations, there are obstacles in military organizations to testing, experimentation, and self-evaluation. Enhancing the capacity for rational analysis requires a widespread educational effort, underlining the importance of a professional military education that imparts norms of empiricism.

In the matter of the B-52 program, as I. B. Holley and other historians have observed, AMC officers had better technical educations than those on the Air Staff. The relationship between technical background and assignment to AMC, an organization created to deal with aerodynamic R&D and aircraft production, is not surprising. It also is not surprising that AMC officers employed empirical premises more consciously to evaluate prospects for technological advance. What is interesting, however, is the way debate and discussion between Air Staff and AMC resulted in a reasonable process for dealing with uncertainty and technological ambiguity. The Air Staff-AMC arguments mitigated the technical weaknesses of Air Staff officers and short-circuited the tendency of AMC officers to focus on marginal technical improvements of existing aircraft types. The case study shows that enhancing the capacity for rational analysis may be achieved by a widespread educational effort, but there is no necessary relation between education and analysis. The competitive relationship between Air Staff and AMC officers achieved the same end-the acquisition of an effective aircraft.

How Does Organizational Structure Effect Acquisition Outcomes?

The central argument of this study is that the key to understanding how to exploit rapidly advancing technology lies in the relationship between organizational structure and outcomes, and the process through which sets of organizations with overlapping jurisdictions interact. The concept of multiorganizational systems introduced earlier is central to dealing with organizational pathologies that are difficult to manage in a single organization. It is important to review here the properties of the concept and its application to understanding the innovation process.

Briefly, two or more organizations acting together create a social system. Over time, member organizations assume particular roles and develop expectations of the others' behavior. The interaction of separate organizations generates a new level of analysis. The multiorganizational system has an identity separate from its members, and many activities of a multiorganizational system cannot be explained simply by examining the behavior of its members. [10]

A critical aspect of multiorganizational systems concerns how actions are coordinated. Coordination is, of course, a problem also for individual organizations and may be approached through the concepts of "tight coupling" and "loose coupling." Loose coupling "allows the sequence of a set of components to be changed, making alternatives available; while tight coupling connects a sequence that is fast moving, no by-passes or alternative channels, and will only work in one fixed order." [11] The two types of coordination are essentially end points on a scale used to describe organizational structure. These end points need not describe real points; groupings of actual organizations will be found at varying positions along the scale. [12]

In a tightly coupled individual organization or multiorganizational system, hierarchical position and formal authority allow superiors to determine the actions of subordinates. The division of labor and specialization within or among organizations anticipates and predicts problems faced, and provides solutions in the form of decision rules, routines, or standard operating procedures. [13] Coupling, in this context, is a causal concept and refers to control of actions leading to prescribed ends. Planning and execution are linked in a linear sequence: goals are enunciated and translated into specific plans and policies, plans are reduced to standard administrative procedures, procedures are then executed by diligent subordinates following clear timetables and receiving the proper feedback. [14] It makes sense to employ a tightly coupled organizational structure to deal with well-defined and well-understood tasks: the relationship between means and ends is understood and performance can be evaluated. Tight coupling is the preferred organizational form for military organizations, as is the peaked hierarchy with authority concentrated in a few persons. One property of such a hierarchy is that it reduces the costs of communication and the amount of time required to transfer instructions, and minimizes the need for coordination. The commander issues orders that are not subject to negotiation or discussion and that are implemented speedily.

Thus, tightly coupled decision systems (1) deal with well-structured problems (or are able to convert ill-structured problems into well-structured ones); (2) have a well-defined way to assign problems to subunits, with an explicit division of attention and labor among subunits; (3) use clear sets of procedures to deal with conflict between subunits; and (4) encompass clear information requirements associated with choices. The programmed decisions appropriate to tightly coupled decision systems require agreement about both preferred outcomes (or ends or values) and appropriate means (or method or tools). [15] When such agreement obtains regarding a task or problem, the solution may be delivered as a detailed prescription that governs the sequence of system responses to a complex task environment. There is no need to develop

106

alternatives, bargain, or negotiate. There is no doubt about goals or ends, and the knowledge and technology needed to achieve the ends are available.

Unfortunately, despite the desirable trait of giving the appearance of certainty, the employment of tightly coupled decision systems is inappropriate for poorly understood and uncertain problems. First, the act of programming a solution to a task is not a guarantee that the program itself is adequate to the task. The unproven and hypothetical character of any solution usually is not raised by the partisans of a plan. Nevertheless, programmed decisions are presented as a normative ideal-as the most legitimate mode of decision-even when programmed decision making is not possible. [16] Second, the range of tasks for which programmed decisions in tightly coupled hierarchies are appropriate is small. [17] As Nobel laureate Herbert A. Simon noted, "Many, perhaps most, of the problems that have to be handled at the middle and high levels in management have not been made amenable to mathematical treatment, and probably never will." [18] Tightly coupled organizational systems may no longer be appropriate to all types of warfare. Waging mobile, high-tempo warfare over large territorial areas (with modem communications equipment and complex weapon systems) may require a more loosely coupled organization structure so that human initiative and flexibility may be combined more effectively with advanced technologies.

In the dynamic environment the military services face today, where organizations confront problems that are neither well defined nor well structured, intelligent choice and action are still possible. However, the pervasiveness of uncertainty places a premium on robust adaptive procedures, instead of on procedures that work only when they mesh with precisely known environments and tasks. Typically, organizations face many uncertainties in dealing with ill-structured tasks: key variables or parameters may be unknown (sometimes referred to as unknown unknowns), the magnitude or importance of known variables may be unknown, or goals and preferences may be unstable. [19] One organizational response to dealing with ill-structured decisions is to negotiate the uncertainty or problems away. For example, business firms form cartels to set minimum prices and divide market shares. [20] On the one hand, this strategy to remove uncertainty may be satisfactory when the relevant organizations have incentives to cooperate and can monitor performance. In World War II, for example, Germany, Great Britain, and the United States tacitly agreed to forego use of chemical weapons against each other. On the other hand, negotiating uncertainty away is unlikely to be successful when the "adversary" is the physical world. No amount of pleading or bargaining by humans can alter laws of physics or chemistry.

Inattention to uncertainty may lead to attenuation of the relationship between means and ends at various points, either through "goal displacement" (i.e., the substitution of means for ends) or through premature programming. Goal displacement usually occurs when the organization's goals cannot be connected directly with its actions. As a consequence, decisions are judged against subordinate goals that can be connected to means, that is, whether a particular rule is followed. The question of whether following the particular rule has any effect on the overall goal is ignored or not asked.

The second implication of ignoring uncertainty-premature programming-refers to the tendency of organizational leaders to adopt a plan and order its implementation as if perfect knowledge and value consensus exist, when-in fact-they do not. Organizational members may agree on factual premises that are wrong. The resulting decisions tend to be self-reinforcing and closed to criticism. The rejection of criticism only covers up existing disagreements within the organization, and the lack of consensus becomes a source of conflict-not a basis for negotiation. Despite the presence of agreement about ends, premature programming leads to self-delusion in the face of repeated error, that is, a refusal to learn from mistakes.

Murphy's Law, "anything that can go wrong, will go wrong," operates in most critical situations. No matter how carefully designed and programmed, organizational components-equipment and people-will fail or violate expectations. When interdependent components are tightly coupled into serial chains, failure can cascade along communication channels, leading to unacceptable results. [21] Hence, recognizing the inevitability of error may be the single most important factor in the design of effective organizations and procedures to foster and enhance innovative technology and concepts.

The loosely coupled-flat hierarchy-organization contrasts sharply with the tightly coupled hierarchy. Because there may be no center of authority in multiorganizational systems (or at a lower level of analysis, in single loosely coupled organizations), they may appear to be disordered. [22] Yet organization and cooperation by people are possible without explicit design or procedures telling people what to do and when. Loosely coupled organizations have roles and task definitions that are not set by a single leader or authority. The components themselves set the tasks; technologies are less clear and participation is fluid. As a consequence, interaction and communication among components occur on the basis of need and not as a result of commands or instructions. [23] Organizational roles of components adjust on the basis of experience, and tasks are established by negotiation. The components participating in negotiation about tasks are determined by the particular character of the issue at hand rather than by the organizational chart. [24]

Loosely coupled organizations have many advantages. [25] Loose coupling avoids placing an insupportable burden of calculation on a central planning mechanism. Because novelty does not disrupt established routines very much, loosely coupled organizations are creative, adaptive, and open to innovation. [26]

In loosely coupled organizations senior decision makers are less vulnerable to manipulation by information providers because links between information and its users are not tight. [27] They also are self-regulating because the stimuli for adaptation and innovation are information generated by experience rather than by a priori demands of planners.

A loosely coupled organization is not necessarily fragmented, that is, in need of central direction; such organizations are a social and cognitive solution to constant environmental change. [28] As already noted, loosely coupled systems have roles and definitions of tasks that are not set by any single authority but by the components themselves. Interaction and communication occur not as a consequence of instruction or command, but on the basis of need. Roles are continuously adjusted on the basis of experience, and tasks are generally established by negotiation. Parties to the bargain are determined by the character of an issue, not the organization chart. [29]

It appears that the introduction of carrier-based aviation into the Navy and the jet-propelled B-52 into the Air Force were managed by multiorganizational systems that had many features of loosely coupled systems. The component organizations defined their tasks and set roles through their mutual interaction. Senior decision makers were less vulnerable to the manipulation of information in their own component organizations, and the stimuli for adaptation and innovation were generated by real-world experience.

Final Comments

The ultimate fate of an emerging military revolution is tied to the performance of the economy, society, and political system. Innovations embody both specific technical knowledge and a particular social, economic, and political context. As the twentieth century closes, the acceleration of knowledge and technology has created more opportunities for invention and implementation of innovations. Yet, the management of programs dedicated to fostering innovation can be hobbled by chance, short time horizons, bureaucratic vetoes, and poor decision-making.

Like the Air Staff and AMC officers of the immediate post-war period, national security decision makers today face complex and risky choices. The growing complexity of these choices underlines the importance of identifying the factors that foster effective leadership and either encourage or limit thinking and deciding. Some analysts call on national security decision makers to look beyond current problems and to propose options and solutions to problems that do not yet exist. [30] Such proposals make unreasonable demands upon the cognitive capacity of decision makers. The B-52 case study reinforces the notion that decisions shaping the future more often result from the ongoing search for solutions to immediate problems. Exhortations to plan a synoptic or comprehensive reassessment of technology, management, and organizational opportunities often lead to rhetoric instead of rationality.

Whether a particular organization is set up to learn affects its ability to compete effectively in the innovation process. The most successful organizations often pursue multiple and parallel approaches to a development problem, thus reducing uncertainty by identifying and eliminating poor options. As argued above, an extensive empirical literature supports this view, and this strategy was essentially followed, although somewhat unwittingly, by the Air Force in B-52 development.

Successful organizations foster efficient feedbacks and successful comparison of multiple options within a context of institutional rules that stress empiricist habits of mind and the role of evidence. Thus, systematic improvements in combat effectiveness may require an institutional setting that allows for examination of many choices and the development of feedback mechanisms (e.g., war games and simulations) to identify and eliminate poor choices. The institutional backdrop also must provide acknowledgment and rewards, allowing officers and enlisted personnel engaged in such work paths to higher status, responsibility, and authority.

As discussed in chapter 3, examining history from the standpoint of multiple levels of analysis permits a more rigorous approach to the question of how and where errors enter the decision process, how and where errors are identified and corrected, and how technical and policy uncertainties affect this identification and correction of errors. In the mid- to late-1940s, the loosely-coupled interaction among Air Force organizations and private firms-like the interactions among Navy organizations concerned with aviation during the interwar period-helped senior decision makers learn about and anticipate errors, problems, and options. Senior decision makers today can consciously create a self-correcting multiorganizational system by setting institutional rules that permit interaction of diverse and independent groups and agencies. Thus, policy makers who wish to foster innovation should devote attention to coordinating the interactions among agencies.

This monograph's analysis points to specific features of links between process and outcome; some of these links are quite different from those long advocated, and they lead to practical lessons that can be implemented by policy makers.

• Efforts to improve the ability to innovate should focus on the interaction of organizations and agencies within a policy jurisdiction rather than on forming a dedicated analytic organization.

• As a technological innovation process proceeds, it is critically important to have independent sources of information and competing analyses to avoid the problem of uncertainty absorption. [31] Policy makers should refrain from the use of efficiency criteria and discourage "streamlining" of individual organizations as a matter of principle.

• Overlap of policy jurisdictions is useful, as it encourages the formation of multiorganizational systems. This perspective supports an argument for balanced authority of Office of Secretary of Defense (OSD) and Joint Staff, and against allowing either the Joint Staff or OSD overwhelming influence over the other. In addition, independent and separate organizations should always be involved in analyzing important policy problems.

• Competition and interaction among offices is necessary to stymie efforts to negotiate away uncertainty.

• Frequent rotation of officers in and out of offices that oversee R&D may have a detrimental effect on the ability of the organization to generate good decisions.

These important and useful conclusions can begin to inform decision making about innovation today. In fact, this monograph's analysis supports the "integrated product and process development" and "integrated product teams" approach to defense acquisition described by Undersecretary of Defense for Acquisition and Technology Paul G. Kaminski in May 1995. [32] The crux of this reform is the creation of an organizational structure that can coordinate and integrate the vastly different views of relevant "functional disciplines"-the contracting firms, the military users, test and logistics communities, and the Office of Secretary of Defense and service program and financial managers-from design through manufacturing and system support.

The research presented here also suggests directions for further study. For instance, examining the composition of multiorganizational systems and the role they played in creating major innovations during the 1950s will illuminate further the factors that structure present and future choices, and reinforce the value of multiorganizational thinking around military revolutions.

Future research also should continue to examine how multiorganizational systems operate and the interrelationship between institutions and organizational and individual behavior. It would be useful to study how a preplanning, programming, and budgeting system set of institutional rules (access to decision makers, definition of decision makers, appropriate arguments and evidence used, factors considered) either constrained or created opportunities for action and how these institutional rules influenced the ability of some technologies to be adopted over others and how operational concepts and technology were implemented.

In closing, this monograph should leave senior decision makers with the admonition that they have no illusions about the ease of implementing organizational reforms capable of fostering innovations. They must also keep in mind three factors that effect the potential of the military to achieve a revolution in combat capability. First, effective military acquisition organization and policies require a description of the situation that is to be acted upon. Such a description contains assumptions about causality and, therefore entails an explanation. But analysts often disagree about the proper factors to examine. This monograph, for example, began with the observation that understanding military revolutions should be represented in terms of five diverse factors, including the relationship between organizational structure and outcomes; questions of knowledge and evidence; the impact of the American constitution and process of government on innovation opportunities; the ideas, behavior, and experience of individuals over time; and the political economy of technological change. Second, policy making presumes that there are solutions to problems and valid goals for which to strive. Yet the relationships among problems, goals, and solutions are not always simple or clear. The proposed solution may not achieve the desired goal because of unexamined second- or higher-order effects, or because it is causally irrelevant to the problem. Achieving the desired goal may not even resolve the original problem. Third, the manner by which a policy solution is implemented has a critical role in its success or failure. Poor implementation may vitiate a program even when all other factors have been anticipated correctly.

Perhaps paradoxically, the most successful innovators may be those sensitive to these myriad opportunities for failure. As Justice Oliver Wendell Holmes Jr. once observed, "Every year, ' if not every day, we have to wager our salvation upon some prophesy based on imperfect knowledge."

Notes

1. This claim is consistent with the experience of the Navy's Special Projects Office (SPO) as it managed the development of the fleet ballistic missile. Although part of SPO's success lay in the use of a new management technique to isolate and shield the organization from political interference. Adm William F. Raborn and Adm Levering Smith were able to create a small multiorganizational system that integrated the technical views of many different sources and groups.

2. However, it should be acknowledged that evaluations of combat sometimes do not settle questions and debates about doctrine or weapons effectiveness.

3. Baruch Fischhoff, "For Those Condemned to Study the Past: Heuristics and Biases in Hindsight." in Judgment Under Uncertainty: Heuristics and Biases, ed. Daniel Kahneman, Paul Slovic, and Amos Tversky (Cambridge: Cambridge University Press, 1982), 341.

4. Imre Lakatos, "Falsification and the Methodology of Scientific Research Programmers," in Criticism and the Growth of Knowledge, ed. Imre Lakatos and Alan Musgrave (Cambridge: Cambridge University Press, 1970).

5. Roberta Woshlstetter, Pearl Harbor: Warning and Decision (Stanford: Stanford University Press, 1962), 401.

6. Michael Polanyi, "The Potential Theory of Adsorption." Science, 13 September 1963, 1013.

7. John E. Steiner, "Jet Aviation Development: A Company Perspective," in The Jet Age: Forty Years of Jet Aviation, ed. Walter E. Boyne and Donald S. Lopez (Washington, D.C.: Smithsonian Institution Press, 1979), 143.

8. For example, until the advent of jet propulsion, wooden aircraft structures were as efficient as their aluminum equivalents. One of the most successful World War II propeller-driven combat aircraft, the de Havilland Mosquito, had a mostly wooden structure. Wood structures were more weight efficient than aluminum. The disadvantages of wooden aircraft structures included poorer response to humidity and heat and greater difficulty of fabrication. In contrast, fabricating aluminum structures required a great deal of expensive electrical energy. Eric Schantzberg. "Ideology and Technical Choice: The Decline of the Wooden Airplane in the United States, 1920-1945," Technology and Culture 35, no. 1 (January 1995): 34--69.

9. Marcelle S. Knaack, Encyclopedia of U.S. Air Force Aircraft and Missile Systems, vol. 2, Post-World War II Bombers 1945-1973 (Washington. D.C.: GPO, 1988), 215-18.

10. Andrew H. Van de Ven, "On the Nature, Formation, and Maintenance of Relations Among Organizations," The Academy of Management Review I, no. 4 (October 1976): 24-36.

11. Eliot A Cohen and John Gooch, Military Misfortunes: The Anatomy of Failure in War (New York: Vintage Books, 1991), 23.

12. This logic of comparison is explicated by Carl G. Hempel, "Typological Methods in the Social Sciences," in Philosophy of the Social Sciences: A Reader, ed. Maurice A. Natanson (New York: Random House, 1963), 213.

13. See Martin Landau, "On the Concept of a Self-Correcting Organization," Public Administration Review 33 no. 6 (November/December 1973): 533-42; William H. Starbuck, "Organizational Growth and Development," in Handbook of Organizations ed. James G. March (Chicago: Rand McNally, 1965).

14. Leonard D. Sayles, "Technological Innovation and the Planning Process," Organizational Dynamics, Summer 1973, 68-80.

15. This type of decision is named programmed because a correct decision is governed by a program or set of rules, as in a computer program. See James D. Thompson and Arthur Tuden, "Strategies, Structures, and Processes of Organizational Decision," in Comparative Studies in Administration ed. James D. Thompson and Arthur Tuden (Pittsburgh: University of Pittsburgh Press, 1959); Martin Landau, "Decision Theory and Comparative Public Administration," Comparative Political Studies, July 1968, 175-95. For March and Simon, "the term 'program' is not intended to connote complete rigidity. The content of the program may be adaptive to a large number of characteristics of the stimulus that initiates it." James G. March and Herbert A Simon, Organizations (New York: Wiley, 1958), 142.

16. David Braybrooke and Charles E. Lindblom, A Strategy of Decision: Policy Evaluation as a Social Process (New York: Free Press, 1963).

17. This issue also has been dealt with at great length in the complementary research of philosopher of science Karl R Popper and (Nobel laureate) economist Friedrich A. von Hayek. Both saw their research during, and before, World War II as efforts to analyze and undercut the desirability of planning philosophies to guide society. For von Hayek, the prime argument in favor of markets was not grounded in ideology; it was based on a recognition of the cognitive limits of central planners. For Popper, the prime philosophical argument against planning centered on epistemology: on the limits of what even a perfect planner could know. See Friedrich A. Hayek, The Road to Serfdom (Chicago: The University of Chicago Press, 1944); Hayek, "The Use of Knowledge in Society," American Economic Review, September 1945. 519-30; Karl R Popper, The Poverty of Historicism (New York: Harper Torchbooks, 1957); Hayek, The Open Society and Its Enemies: The Spell of Plato (Princeton: Princeton University Press, 1971); Hayek, The Open Society and Its Enemies: The High Tide of Prophecy: Hegel, Marx, and the Aftermath (Princeton: Princeton University Press, 1971).

18. Herbert A Simon, The Shape of Automation for Men and Management (New York: Harper & Row, 1965), 58-59.

19. Allen Newell, Heuristic Programming: Ill-Structured Problems," in Papers in Operations Research. vol. 2, ed. Julius S. Aronofsky (New York: John Wiley and Sons, 1969); Walter R. Reitman, Heuristic Decision Procedures, Open Constraints, and the Structure of Ill-Structured Problems," in Human Judgments and Optimality ed. Maynard W. Shelly, III and Glenn L. Bryan (New York: John Wiley and Sons, 1964); Herbert A Simon, "'The Structure of Ill-Structured Problems," Artificial Intelligence, 1973, 181-201.

20. Organizations devise and negotiate an environment so as to eliminate uncertainty. Rather than treat the environment as exogenous and to be predicted, they seek ways to make it controllable." Cyert and March, A Behavioral Theory of the Firm, 120.

21. Martin Landau, "On Multiorganizational Systems in Public Administration," Journal of Public Administration Research and Theory, January 1991, 5-18.

22. Karl E. Weick, "Educational Organizations as Loosely Coupled Systems," Administrative Science Quarterly, 1976, 1-19.

23. Weick notes that a "loosely coupled system is not a flawed system. It is a social and cognitive solution to constant environmental change." Karl E. Weick, "Sources of Order in Underorganized Systems: Themes in Recent Organizational Theory," in Organizational Theory and Inquiry, ed. Yolanda S. Lincoln (Beverly Hills: Sage Publications, 1985), 121.

24. Landau, "On Multiorganizational Systems in Public Administration."

25. One obstacle to implementing loosely coupled organization in established formal organizations concerns the apportionment of credit (and reward) for successful performance. Instead of the senior leadership claiming credit for success (as occurs in tightly coupled hierarchies), the "team" assembled to deal with a particular task receives the credit. I am indebted to Jacob Neufeld for identifying this issue.

26. The adaptability of loosely coupled military organizations in wartime may offer a model for domestic agency programs (e.g., urban renewal or revitalization). Such programs need the ability to adapt quickly to a particular city's needs by identifying and creating links with appropriate city groups. Wilson cites Martha Derthick who noted that when Congress sets up new programs very quickly, it implicitly requires that agencies become capable of responding-"capable of devising new routines or altering new ones very quickly-[these qualities] are rarely found in large formal organizations." Wilson adds that "government agencies are far less flexible than formal organizations generally." During the Persian Gulf War, US Central Command (USCENTCOM) and US Central Command Tactical Air Force (USCENTAF) were very flexible: so flexible that a completely ad hoc organization provided command and control for the air campaign. See James Q. Wilson, Bureaucracy: What Government Agencies Do and Why They Do It (New York: Basic Books. 1989), 368.

27. M. A Feldman and James G. March, "Information in Organizations as Signal and Symbol," Administrative Science Quarterly, 1981, 171-86; Jacob A. Stockfisch, Incentives and Information Quality in Defense Management, R-1827-ARPA (Santa Monica: RAND Corporation, 1976).

28. Weick, "Sources of Order in Underorganized Systems: Themes in Recent Organizational Theory," 131.

29. Landau, "On Multiorganizational Systems in Public Administration." 30. See Paul Bracken, "The Military After Next," The Washington Quarterly 16, no. 4 (Autumn 1993): 157-74. 31. The costs of uncertainty absorption can be great. For example, administrative and congressional sources say the DOD may have spent billions of dollars to counter false Soviet threats: threats that Central Intelligence Agency (CIA) officials forwarded to the highest governmental levels without caveats about their possible tainted origins. See Walter Pincus, "CIA Passed Bogus News to Presidents," The Washington Post, 31 October 1995, AI, AID; Walter Pincus and R. Jeffrey Smith, "CIA Data Skewed Cost for Defense," The Washington Post, 2 November 1995, AI, A14.

32. William J. Perry, secretary of defense, memorandum, subject: Use of Integrated Product and Process Development and Integrated Product Teams in DOD Acquisition, 10 May 1995.

APPENDIX
XB-52 Program Select Senior Personnel

Person/Rank	Air Staff Office/Tour	AMC Office/Tour	Other Office/ Tour
Maj Gen Franklin O. Carroll	Office, AC/AS-4, Materiel (6/1947-10/1947)	Assistant Deputy Commander, Engineering (1/1947-/1947): Director. R&D (10/1947-0/1949)	
Gen Benjamin W. Chidlaw		Deputy Commander, Operations (&/1945-10/1947): Deputy Commander (10/1947-9/1949): Commander (9/1949- 7/1951)	
Lt Gen Howard A. Craig	DCS/M (10/1947-/1949)		
Maj Gen Laurence C. Cragie	Director R&D, Office (10/1947-9/1948) (11/1944-10/1947)	Chief Engineering Division (1947-1949)	
Maj Gen Alden R Crawford	Chief, Research & Engineering, Office (1947-1949)	Chief, Engineering division (1947-1949)	
Col J. S. Holtoner	Chief, Aircraft Branch, Office. DCS/M (1947-5/1951)	(Early program on guided missiles) (1946-1947)	
Col Clarence S. Irvine	Four-engine bomber program) (Late 1930s-1943): Special Assistant to, Aircraft Production (early Strategic Air 1943-1944); Chief. Very Heavy Bomber Command CINCSAC) Program (1945-1946)		Assistant to Commander in Chief Strategic Air Command (CINCSAC 1947)
Maj Gen Curtis E. LeMay	DC/AS for R&D (12/1945-10/1947)		
Brig Gen Alfred R Maxwell	Chief, Requirements Division. AS/ AS-3. Operations (8/1945-7/1947)		
Gen Joseph T. McNarney		Commander (10/1947-9/1949)	
Lt Gen Louris Norstad	AC/AS-5, Plans (6/1946-10/1947): DCS/O (10/1947-1950)		
Person/Rank	Air Staff Office/Tour	AMC Office/Tour	Other Office/ Tour
Maj Gen Edward E. Partridge	AC/AS-3 Operations and Training (1/1946-10/1947; Director, Training & Requirements, Office, DCS/O (10/1947-10/1948)		
Brig Gen Thomas S. Power	Deputy AC/AS Operations: Office, DSC/O (9/1946-6/1948)		Vice Commander, Strategic Air Command (SAC) (10/1948-/1954)
Maj Gen Edward M. Power	AC/As-4, Material (12/1945-10/1947); Assistant DCS/M (10/1947-1949)		
Brig Gen Donald L. Putt	Director, R&D, Office, DCS/M (2/1948-1951)	Chief, Engineering Division (1945-1/1948)	
Maj Gen Frederic H. Smith, Jr.	Special Organizational Planning Group (12/1945-1946); Assistant for Programming, Office, DCS/O (10/1947-8/1950)		Chief of Staff, SAC (4/1946-1947)

Gen Carl A. Spaatz	Commander, AAF (12/1945-9/1947); Chief of Staff (9/1947-4/1948);		
W. Stuart Symington	Secretary of Air Force (9/1947-4/1950		
Lt Gen Nathan F. Twining	Vice Chief of Staff (10/1950-6/1953)	Commander (12/1945-10/1947)	
Gen Hoyt Vandenberg	Vice Chief of Staff (10/1947-4/1948); Chief of Staff (5/1948-6/1953)		
Lt Col Henry E. "Pete" Warden		Chief, Bombardment Branch, Engineering Division (1947-1949)